SpringerBriefs in Applied Sciences and Technology

For further volumes:
http://www.springer.com/series/8884

This volume collects selected topical entries from the *Encyclopedia of Sustainability Science and Technology* (ESST). ESST addresses the grand challenges for science and engineering today. It provides unprecedented, peer-reviewed coverage of sustainability science and technology with contributions from nearly 1,000 of the world's leading scientists and engineers, who write on more than 600 separate topics in 38 sections. ESST establishes a foundation for the research, engineering, and economics supporting the many sustainability and policy evaluations being performed in institutions worldwide.

Alfons Buekens

Incineration Technologies

 Springer

Alfons Buekens
Vrije Universiteit Brussel, VUB
Brussels, Belgium
and Zhejiang University
Hangzhou, China

The contents of this book first appeared as part of the Encyclopedia of Sustainability Science and Technology edited by Robert A. Meyers, originally published by Springer Science+Business Media New York in 2012.

ISSN 2191-530X   ISSN 2191-5318 (electronic)
ISBN 978-1-4614-5751-0   ISBN 978-1-4614-5752-7 (eBook)
DOI 10.1007/978-1-4614-5752-7
Springer New York Heidelberg Dordrecht London

Library of Congress Control Number: 2012954265

Printed on acid-free paper

Springer is part of Springer Science+Business Media (www.springer.com)

Contents

Glossary

Air equivalence ratio	Also, air ratio or air factor, (λ or k), is ratio of actual air supply to the theoretical (stoichiometric) requirements for complete combustion.
Combustion residues	Ash remaining after combustion and consisting of bottom-ash or clinker, and of fly ash, entrained by flue gas and eventually separated. Chemical neutralization of flue gas also yields salts, by reaction of acid gas components with basic additives.
Emissions	Output of pollutants through the stack (= guided emissions), to a minor extent also as diffuse emission, e.g., from waste pit, evaporation of spills, spreading of fly ash, and outgoing leaks.
Gasification	Partial combustion generating flammable gas and conducted with deficiency of air in various reactor types.
Higher heating value (HHV)	Amount of heat produced by complete combustion of a specific unit amount of fuel in oxygen.
Immission	Added atmospheric concentrations attributed to specific sources, e.g., an incinerator plant, and markedly varying with atmospheric conditions. Immissions are modeled on a basis of (a) emissions, (b) their dispersion, and (c) according to variable atmospheric conditions (wind direction and speed, atmospheric stability).
Municipal solid waste (MSW)	Waste produced in a city and collected by the municipality.
Pyrolysis	Thermochemical decomposition of organic material in the absence of oxygen, yielding gaseous (pyrolysis gas), condensable (tar), and solid products (char).
Refuse-derived fuel (RDF)	Fuel from waste, produced by mechanical processing, (possibly biological), drying, and possibly densification.
Waste-to-energy (WtE)	Incineration process in which solid waste is converted into thermal energy to generate steam that drives turbines for electricity generators (http://www.businessdictionary.com/definition/waste-to-energy.html).

Definition of the Subject

Waste incineration is the art of completely combusting waste, while maintaining or reducing emission levels below current emission standards and, when possible, recovering energy, as well as eventual combustion residues. Essential features are as follows: achieving a deep reduction in waste volume; obtaining a compact and sterile residue, yet treating a voluminous flow of flue gas while deeply eliminating a wide array of pollutants.

Destruction by fire is almost as old as humanity. Incineration was systematically applied at some locations, both in England and the USA, from the second half of the nineteenth century [1–4]. Furnaces widely differed in conception, yet were still poked and de-ashed manually. A successful furnace design was the cell furnace, composed of a series of juxtaposed combustion cells with a fixed grate, or also with two superposed retractable grates [4–6]. In 1895, the first large continental incinerator was mounted in Hamburg [7] after traditional export to the countryside of *municipal solid waste* (MSW) was jeopardized by an outbreak of cholera.

The technology was strongly inspired by that of coal firing: *mechanical grate* stokers developed from the 1920s and 1930s were continuously improved to suit the special requirements of firing waste and distributing primary air, while cooling the grate bars [4, 8]. After World War II, *fluidized bed* techniques were introduced mainly in the Nordic countries, where MSW was co-fired together with forest products and residues from pulp and paper industry, and also in Japan, where the suitability of fluidized bed combustors for one- or two-shift operation was valued [9–11]. *Slagging operation*, with tapping of molten residue, remained unusual until the end of the twentieth century; then it became mandatory in Japan to melt fly ash and destroy its organic contents, while either volatilizing or immobilizing its heavy metal content by conversion into a glassy state (vitrification) [12]. A search on "melting" yields more than 130 different processes, as proposed by numerous Japanese corporations [13].

Gasification of waste, a partial combustion conducted with deficiency of air, yields flammable gas, suitable as cleaned gaseous fuel or even for driving engines or turbines [9, 13–16]. This thermal conversion method is mainly apt for high-calorific waste, the complete combustion of which is difficult to control otherwise. Wood waste has been proposed as a decentralized source of heat and power [17, 18].

Pyrolysis, or thermal decomposition of waste [9, 13], may be suitable for specific waste, such as plastics [19], rubber, sewage sludge [20], or wood. These different thermal processes are not to be recommended for general waste, since their process complexity is higher and their availability, hence, lower, whereas most of the advantages claimed often failed to realize [21, 22]. Selecting unproven technology is probably about the worst possible decision in waste management.

Waste varies erratically in *composition* and *properties* and these greatly influence the selection of incinerator furnaces, heat recovery, and flue gas cleaning. Important *waste characteristics* are those determined by *proximate analysis* (moisture, ash, combustibles, subdivided further into fixed carbon and volatile matter) and *elementary analysis* (C, H, N, S, Cl... + O, by difference from 100%) of the combustible fraction. Moist refuse is difficult to ignite. Ash content confines the reduction in weight achievable and determines the burden of residue extraction; important is the composition of ash and its softening and melting behavior at high temperature. *Volatile matter* will lead to flaming combustion and *fixed carbon* to glowing combustion, each of these two modes showing their specific demands. Data from these analyses also allow establishing the necessary material balances, as well as estimating the *higher heating value* (HHV).

Combustion of solid waste proceeds in successive steps as schematically represented in Table 1.

In an incinerator furnace, these successive steps may well proceed in parallel and overlap partly. A combination of chemical reactor engineering and of computer fluid dynamics (CFD) may be used in modeling both physical (flow, mass, and heat transfer) and chemical phenomena (combustion, a complex chemical process, proceeding over numerous reactions involving a large number of intermediates, such as free radicals and ions). Cfr. Furnaces, Their Duties, Peripherals, Operation, Design and Control.

Combustion is never *entirely* complete, even though – thermodynamically – equilibrium approach could come close to unity. In practice, when most combustibles are burned, the rate of heat generation drops, temperature falls and combustion slows down and eventually stops. The reason for further completing combustion is strictly environmental: *Products of incomplete combustion* (PICs) are major atmospheric pollutants and responsible for reduced visibility, photochemical smog, as well as *soot* or *black carbon* formation. PIC's scientific and health aspects are at the heart of dedicated biannual PIC conferences [23]. Cfr. Post-combustion, Dioxins.

Table 1 Successive steps in the combustion of waste

Step	Drying	Pyrolysis	Gasification	Combustion
Evolving to the gas phase	Water vapor	Volatile matter	Carbon monoxide, hydrogen, methane	Carbon dioxide, water vapor
Residue	Dry waste	Char and ash	Ash	Ash

Coarse *combustion residues* (USA: *bottom ash* or slag; UK: *clinker*) are the principal residues of MSW incineration. After removal of unburned material and metals, these may be weathered, graded, and recycled as an aggregate material in sub-road construction and embankments [24–26] Cfr. Residues.

Fly ash is separated by flue gas dust filtration. It is considered *hazardous*, because it accumulates volatilized heavy metals (e.g., Hg, Cd, Pb, and Zn), as well as PICs, some of which are semi-volatile, such as *polycyclic aromatic hydrocarbons* (PAHs) and *dioxins* i.e., polychlorinated dibenzo-p-dioxins (PCDD) and dibenzofurans (PCDF), listed and targeted for removal and destruction by the *Stockholm Convention* on *principal organic pollutants* (POPs). Thermal processes have been applied to detoxify such residues [27–29], yet this treatment is expensive.

Incinerator furnaces. The selection of *furnace types* mainly depends on the characteristics of the waste and the strategies followed to feed the waste to be fired, to contact it with combustion air and to extract the combustion residues from the furnace. Construction of furnaces has evolved mainly empirically, with trial and error as the main method. Tremendous progress made in combustion sciences has started to see some more applications in incinerator design and operation.

Incineration requires sufficient *combustion air*, as well as suitable levels of the three T's, i.e., *T*emperature, residence *T*ime, and *T*urbulence. Turbulence is required to sustain the required macro- and micro-scale mixing to bring together combustibles and air oxygen. Conditions during incineration vary according to the technology employed and the characteristics of the waste fired. Some combustors feature active heat and mass transfer so that combustion takes place much faster, e.g., in vortex or fluidized bed burning. These require, however, size-reduced waste, i.e., preliminary shredding and grading, so that the residence time provided allows either complete burnout even of the largest particles or their recycling after separation.

Temperature ranges from as low as about 750°C (bed temperature of fluidized bed combustion) to more than 1,200°C (destruction of hazardous waste, such as PCBs, slagging operation). High temperatures are only moderately beneficial, for de-mixing of fuel, and oxygen controls combustion rates. *Pressure* is often slightly below atmospheric, to restrict the emanation of combustion products, smoke, and grit. *Residence time* at high temperature is only few seconds (generally 2–3 s) for flue gas. Solid waste and its combustion residue have a much longer residence time, from about a minute in fluidized bed combustion (time required to dry, heat and burnout the ash) to typically half an hour on a mechanical grate; yet, much depends on the time required for drying and heating. After ignition, combustion of volatile matter proceeds rapidly, but burnout of fixed carbon may take time in case of diffusion-controlled combustion, e.g., of ash-occluded carbon.

Some codes prescribe minimum values for temperature and time (e.g., 850°C for 2 s), or they limit the amount of products of incomplete combustion in flue gas (*carbon monoxide*, CO; *total organic carbon*, TOC) and carbon in residues.

Combustion air is supplied to the furnace with several purposes: *primary air* activates the fire bringing oxygen to the reaction surroundings, whereas *secondary air* (also termed over-fire air) is injected at high speed (typically 100 m/s) to induce mixing, as far as its momentum reaches. Increasing primary airflow accelerates combustion until a point at which higher cooling supersedes this stimulating effect. Air may also be used for cooling furnace walls and mechanical grates. Since several decades, water-cooled grates are also in use.

Incinerators are thermal units: liberating more combustion heat also requires supplemental combustion air. A simple rule of thumb states that this amount is directly proportional to the higher heating value, whatever the fuel fired (gas, oil, coal, or garbage of any kind).

In order to obtain complete combustion, it is essential that an adequate amount of air oxygen is supplied. The *air equivalence ratio* indicates the actual air supply, compared to the theoretical, stoichiometric requirements for complete combustion. The difference, the excess air, merely cools the flame and inflates the volume of gas to be cleaned. Better mixing of fuel and air allow operating at lower air equivalence ratios.

One Mg (metric tonne) of MSW typically generates some 5,000–6,000 m^3 of flue gas! Flue gas flow varies proportionally with both the higher heating value and with the amount of excess air. In numerous plants, the uncontrolled entrance of air leaking into furnace and flues seriously inflates the volume of gas to be cleaned.

During waste combustion, the spatial distribution of flames (formed by combustion of volatile matter) is unpredictable and hence results in erratically active combustion zones, showing oxygen deficiency and less active zones, where oxygen requirements are much less and oxygen plentiful. This results in a complex pattern of oxygen-rich and oxygen-deficient strands that should be mixed intimately in order to reach complete combustion.

Combustion air may also be replaced by oxygen-enriched air, or even by pure oxygen, in order to improve and accelerate combustion. Such practice markedly reduces the volume of flue gas, yet it considerably adds to operating expense and is limited to exceptional cases, such as gasification by means of oxygen/steam mixtures to convert waste into *synthesis gas*.

Municipal solid waste incineration evolved into a complex plant, as represented in Table 2.

MSW storage generally takes place in a deep pit, made of impervious concrete. Storage bridges the gaps between the schedules of collection rounds and continuous firing. A traveling crane allows mixing waste of different origins, stacking waste against the bunker wall, and feeding it into the hopper on top of the load shaft. In the USA, storage floors are in widespread use.

Boiler plant. The heat from flue gas is transferred to the water, boiling in vertical pipe panels, constituting the boiler and organized around the combustion chamber (for an integrated boiler) and in successive vertical passes of the flue gas. An alternative is to suspend boiler tube panels in a horizontal flue gas channel. The resulting medium pressure steam (at 2–4.5 MPa) is superheated in case the steam is used for power generation. At lower temperature, the flue gas preheats the boiler feedwater in an economizer, and possibly the combustion air in a flue gas/air heat exchanger.

Table 2 Composition of current municipal solid waste incinerator plant

Unit	Function	Potential problems
Storage	Bridging the gaps between delivery and firing of MSW	Dust, smells, fires
Crane	Traveling crane to mix and load MSW into a hopper	Mechanical
Hopper	Receiving mixed MSW from the storage bunker	Bridging
Valve	Sliding valve to close the furnace	Mechanical
Shaft	Junction with the combustion chamber	Air infiltration
Furnace	Combustion chamber	Refractory spalling or slagging
Grate	Mechanical grate, supporting, conveying, and poking MSW	Wear, clogging
Burner	Start up combustion, maintain temperature if required	
Boiler	Recovers the heat of combustion from flue gas	Fouling, corrosion, erosion
Dust collection	Separate the bulk of the dust from flue gas	
Scrubber	Acid gas neutralization	Corrosion, erosion, deposits

Flue Gas Cleaning

Incineration was once a source of smoke and grit. These have been mastered by improved combustion conditions and deep removal of fine dust: once reduced to below 100 mg/Nm^3 by an electrostatic precipitator, the flue gas becomes invisible, a feature that still satisfied the public in the 1950s and 1960s.

The German emission code *TA-Luft* (Technische Anleitung zur Reinhaltung der Luft), already in its first version (1974) specified emission levels requiring acid gas levels to be reduced. Since, cleaning the flue gas from waste incineration has steadily become more complex and comprehensive, throughout the 1980s and 1990s. Tables 3 and 4 show both the extent of this gas cleaning duty and the frenetic evolution of these codes in time. The European Union also promulgated successive directives on waste incineration (last directive – Directive 2000/76/EC of the European Parliament and of the Council of 4 December 2000 on the incineration of waste) and prepared codes of good practice (BREF reports: BREF stands for *B*AT *Ref*erence Document; BAT = *B*est *A*vailable *T*echnology). Other countries (the USA, Japan, and China) use distinct sets of emission Codes and reporting procedures.

For a good understanding of emission limit values, it is of interest to look at the ratio:

$$\text{Reduction ratio} = (\text{Input value})/(\text{Output value})$$

The reduction efficiencies required (in Table 3) of 95% and 99.9% respectively convert into a reduction ratio of 20 and 1,000, respectively. The first two numbers

Table 3 Raw gas concentration, emissions, and required separation rate of flue gas cleaning devices (Adapted from [30])

	Raw gas concentration (mg/Nm^3, dry)	Emission limit value (mg/Nm^3, dry)	Required reduction rate(%)
Dust	2,000–10,000[a]	10	99.9
HCl	400–1,500	10	>99
HF	2–20	1	95
SO$_2$	200–800	50	94
NO$_x$ (as NO$_2$)	200–400	200	50
Hg	0.3–0.8	0.05	88
Cd, Tl	3–12	0.05	>99.5
Dioxins and furans	1–10 (in ng I-TEQ/Nm^3)	0.1 (in ng I-TEQ/Nm^3)	99

[a]For fluid bed plant these figures are typically 10,000–50,000 mg/Nm^3, dry

Table 4 Some milestones in the evolution of emission limit values (Germany and the European Union)

Compound	TA-Luft Germany, 1974	EU directive 89/369	17. BImSchV[b,c] Germany, 1990	Unit
Dust	100	30	10 (30)	mg/Nm^3
HCl	100	50	10 (60)	mg/Nm^3
HF	5	2	1 (4)	mg/Nm^3
SO$_2$	–	300	50 (200)	mg/Nm^3
NO$_x$	–	–	200 (400)	mg/Nm^3
TOC			10 (20)	mg/Nm^3
CO			50 (100)	mg/Nm^3
Heavy metals,[a]				
• Class I	20	0.2	0.5	mg/Nm^3
• Class I + II	50	0.2	0.05	mg/Nm^3
• Class I + II + III	75			mg/Nm^3
Dioxins and furans	–	0.1	0.1	ng TE/Nm^3

[a]The comparison is distorted by changes in the definition of various classes
[b]17. BImSchV Ausfertigungsdatum: 23.11.1990. Complete citation: "Verordnung über die Verbrennung und die Mitverbrennung von Abfällen in der Fassung der Bekanntmachung vom 14. August 2003 (BGBl. I S. 1633), die durch Artikel 2 der Verordnung vom 27. Januar 2009 (BGBl. I S. 129) geändert worden ist" Cfr.: http://www.gesetze-im-internet.de/bundesrecht/bimschv_17/gesamt.pdf
[c]The 17th BundesImmissionsSchutzVerordnung gives values for a daily average, as well as for a 30-min average, the latter in parentheses

seem deceptively nearby, separated only by 4.9%; the second, the reduction ratios, come closer to the efforts really required in flue gas cleaning, which differ by a factor of 50!

The present emission values are monitored and registered continuously. Some parameters (O$_2$, CO$_2$, H$_2$O) remain rather constant; others are more variable (HCl) or are marked by a continuous value, spiked by peaks (CO, TOC). Dioxins cannot be monitored continuously, yet may be sampled continuously and checked on a weekly or biweekly basis.

Dust Collection

Traditionally, cyclones or *electrostatic precipitators* (ESPs) featuring 2, 3, or 4 consecutive fields were arresting the evolving grit and dust, with an efficiency approaching unity according to an exponential curve. As a consequence, it is increasingly difficult to collect the last particles. Important parameters are the size and electric resistivity of the particles to be collected, as well as their behavior (cake severance or re-entrainment) at the moment of rapping the collection electrodes. Moreover, ESPs operating at temperatures substantially above 200°C were found to generate considerable amounts of dioxins.

Current codes require retention also of the small particles around a micrometer in diameter: even though they correspond to only minor amounts when expressed in mass units (mg/Nm^3), they represent relatively large numbers of particles, strongly enriched in pollutants. *Baghouse filters* (BHFs) are capable of efficiently collecting these particles; moreover, they accumulate a layer of basic substances (injected lime, fly ash) that react with acid gases, such as HCl, SO$_2$, and HF, from the flue gas and adsorb some semi-volatiles.

Neutralization of Acid Gases

Historically, several solutions have been developed to the acid gas problem: wet scrubbing, dry scrubbing, semi-wet scrubbing, and semi-dry scrubbing. Generally, to neutralize these acid gases, hydrated lime is injected into the flue gas (*dry, semi-dry*, i.e., after further moistening the flue gas, and *semi-wet scrubbing*, using water slurries of lime). Wet scrubbing is even more efficient, since the principal acid gas, HCl, is eminently water soluble; yet it is also more complex and capital intensive because of the necessity of maintaining a water circuit and treating the resulting wastewater, removing organic compounds as well as sludge and heavy metals. Moreover, wet scrubbing generally is conducted in two steps: in the first, acid scrubbing (pH 0–2) the bulk of HCl is removed and SO$_2$ follows in the second step, conducted under mild acid or basic conditions (pH 6–8). However, unless the scrubbed flue gas is reheated, wet scrubbing generates a visible plume of condensing water droplets, with its concomitant negative psychological impact. Still, deeper cooling of scrubber liquors also deepens the removal of virtually all pollutants, including mercury and the various PICs (Table 5).

Table 5 Typical stoichiometric factors applied in flue gas cleaning (acid gas neutralization) [30]

Flue gas cleaning	Semi-dry	Semi-wet	Wet
Range (as cited)	2.4 to >3	2.2–3.0	1.1–1.4

Products of Incomplete Combustion – Organic Semi-volatile Micropollutants

In principle, ensuring steady, high-quality combustion and avoiding all combustion upsets should control *products of incomplete combustion* or *PICs*. The latter relate to large masses burning together rapidly and to poor mixing of the intrinsically heterogeneous input.

Much attention has been given to organic *semi-volatile micropollutants* (PAHs, dioxins) that occur in only minute amounts ($\mu g/Nm^3$ and even ng/Nm^3), yet are persistent and bio-accumulating. These compounds are largely removed (>99%) by baghouse filters, after their adsorption onto fine activated carbon particles (typically injected at a dosage of 50–200 mg/Nm^3) or else provided as a fixed adsorption bed.

As an alternative, they are oxidized by means of suitable DeNOx-catalysts, active already at a very low temperature (200°C). A number of preventive measures also allow reducing the formation of PAHs and dioxins (cfr. Dioxins).

Nitrogen Oxides

Nitrogen oxides are formed during combustion, by means of complex free radical and even ionic mechanisms. NO is formed at high temperature and eventually emitted into the atmosphere. In air, slow oxidation of NO takes place, forming strongly oxidizing NO_2. Together with NO, this NO_2 forms an atmospheric oxidizing-reducing system, responsible for the formation of photochemical smog (smog = smoke + fog) and haze. Nitrogen oxides are hence termed "NOx" (NO + NO_2) and generally expressed as their NO_2 equivalent.

NOx in flue gas derives from mainly two sources: the incineration of organic N-compounds (fuel NOx) and incineration at high temperature, e.g., in cement kilns or during slagging operation (thermal NOx and also prompt NOx).

When desirable or required by codes, such NOx can be thermally (*selective non-catalytic reduction*, SNCR) or catalytically reduced (*selective catalytic reduction*, SCR) by means of suitable reducing agents, such as ammonia, urea, amines (N-compounds), hydrocarbons (reburning), and others. Thermal reduction is only possible in a high temperature window, of 760–1,000°C. Catalytic reduction is active already at much lower temperatures, typically 250–450°C.

Another nitrogen oxide is known as nitrous oxide (N_2O), or laughing gas. It forms preferentially at medium-low combustion temperature, such as the fluidized bed combustion of sewage sludge, and during reduction of the conventional NOx. It is a naturally occurring regulator of stratospheric ozone and a major greenhouse gas and air pollutant.

Heat Recovery

Heat recovery has always been central in incineration, and at times waste was regarded as free fuel, yet heat recovery is generally uneconomic in small plants. Some plants incorporate captive uses for the heat produced, e.g., by being linked to district heating systems (Denmark, Sweden) or integrated with civic centers, featuring swimming pools, sauna, and hot baths (Japan), yet generally it is difficult to market the heat produced, so that power generation emerges as a last resort, albeit at limited efficiency. Moreover, the presence of boiler and turbo-generator inflates plant downtime.

Sensible heat is difficult to recover from flue gas, since it is both fouling and corrosive. These limit the possible operating pressure of a *waste heat boiler* (consecutive to the incinerator furnace) or of an integrated boiler, with the furnace fully integrated into its boiler structure (used for highly calorific waste only). Low boiler pressure limits the possible conversion efficiency of steam energy into power. Typically, such conversion efficiency into power is only 16–24%, based on the HHV of waste compared to better than 40% for large fossil fuel–fired thermal power plants (cfr. Heat Recovery).

Co-firing. Waste can also be co-fired in non-dedicated thermal units, such as thermal power plants, cement or limekilns, and in large industrial boilers. Not all waste is suitable, though, because of both combustion and gas cleaning considerations. Table 6 lists the typical requirements for co-firing in cement kilns (cfr. Co-firing of Waste or of RDF, Thermal Power Plants, Cement and Lime Kilns).

Table 6 Some specifications for RDF to be fired in cement kilns [31]

Element	Typical value (ppm)	Limit value (ppm)	Hazardous waste* (ppm)
As	9	20	300
Be	0.4	2	50
Cd	3	5	(+ Tl) 90
Co	8	15	300
Cr	40	120	3,000
Cu	100	150	3,000
Hg	0.6	1	5
Mn	50	150	2,500
Ni	50	100	2,000
Pb	50	100	2,000
Sb	25	60	150
Se	5	10	80
Sn	10	40	1,500
Te	5	20	80
Tl	1	2	
V	10	20	1,500
Zn	n.a.	n.a.	15,000

Source: Reference [31]

Cost and Plant Availability

Incineration is a technically complex and expensive operation. In the European Union, an all-in cost factor for MSW incineration is ca. 100 €/Mg (1 Mg = 1 metric tonne). In Japan, this cost is about three times higher. Internal comparison is difficult, because of highly variable cost factors corresponding to buildings, other infrastructure and, in Japan, land.

Plant availability typically ranges from 84% to 92%, the latter catering for an annual shutdown, the former accounting for repeated and unscheduled stops. Availability heavily depends on the quality of plant management and maintenance.

Public Acceptance

For a variety of reasons, environmentalists have fought incineration as a waste management option: it is not natural (like composting), destroys recyclables, and generates toxic compounds. This opposition is often termed the *not in my backyard* (NIMBY) syndrome and is sometimes counterproductive to the development of adequate solutions on a sound technical and economic basis. (cfr. Public image of Incineration).

Whatever the quality or foundation of the arguments against incineration, the design and operating standards have been much further improved over recent years and today's incinerator emission standards are probably the toughest in industry.

Chapter 1
Introduction

This introduction situates the position of waste incineration in a wider scope of waste management. Traditional waste management was limited to the three options: landfill, composting, and incineration. Landfill was suitable for reclaiming low-value lowlands or restoring the landscape affected by mines and quarries (sand, gravel, clay). Some large cities (e.g., London!) used MSW to fill lowlands, as well as empty quarries of sand, gravel, or clay, to build artificial islands (Tokyo), or even dumped MSW into the sea (New York, Istanbul). Lack of preliminary hydrogeological study and of adequate barriers to contain the leachate has led at times to serious contamination of groundwater. Moreover, landfills are responsible for important high greenhouse gas emissions (methane, carbon dioxide). Composting is still applied nowadays on selectively collected organic fractions; raw MSW yields an unacceptable quality of compost, due to the presence of heavy metals. Incineration has been widely practiced in densely populated regions, where land is at a premium (large municipalities, Japan, Switzerland) and volume reduction primordial.

The 1970s introduced numerous new concepts into waste management, such as the concept of special (Germany), poisonous (England), toxic (Belgium), chemical (the Netherlands) or otherwise hazardous waste (USA, OECD), producer responsibility, the Polluter Pays principle, and mandatory recycling. In the early 1970s, the European Union declared itself competent in environmental matters and the first Framework Directive on Waste (1975) specified the necessity of appointing authorities responsible for waste management, granting licenses, and inspecting waste processing premises. A number of waste streams received particular attention, e.g., hazardous waste, PCBs, waste oil, and packaging. Industrialized countries were

---

This chapter was originally published as part of the Encyclopedia of Sustainability Science and Technology edited by Robert A. Meyers. doi:10.1007/978-1-4419-0851-3

A. Buekens, *Incineration Technologies*, SpringerBriefs in Applied Sciences and Technology, DOI 10.1007/978-1-4614-5752-7_1,
© Springer Science+Business Media New York 2013

repeatedly confronted with waste scandals; industrial and hazardous waste infrastructure was set up step by step and became a booming business. The lowest possible cost disposal was gradually replaced by high-tech, high-cost options. This transition was smoothened through subsidies supporting the options preferred by government and through levies penalizing low-cost landfill. Waste management was borne by the public sector, the private sector, or by public-private initiatives.

According to the Ladder of Lansink (after Dr. Ad Lansink who is a Dutch politician famous for proposing this waste management hierarchy in the Tweede Kamer [Dutch Parliament] in 1979), the generation of waste should in the first place be either prevented or reduced. Next options are reuse and recycle. Lower-ranking options are incineration (preferably with heat recovery), and landfill. Waste management is a legislation-driven business. In several EU countries and in Switzerland, the landfill option is increasingly restricted, so that combustible waste can no longer be landfilled.

Developing countries are often confronted with fast urbanization, so that public services cannot follow demand. Moreover, waste is rich in organics and barely combustible. Large Chinese cities at present are entirely surrounded by a girdle of landfills, polluting groundwater and generating hazardous fermentation gas. Incineration makes rapid progresses, using imported as well as adapted self-developed technology. The severe acute respiratory syndrome (SARS) was material in promoting incineration, in particular for hospital waste. As in numerous developing countries, Chinese MSW is still barely combustible, without resorting to auxiliary fuel!

Chapter 2
Evaluation of Waste Incineration

In brief, waste incineration can be summarized as follows.

Advantages

- It eliminates objectionable and hazardous properties, such as being flammable, infectious, explosive, toxic, or persistent.
- Putrescible matter is sterilized and destroyed. Pathogen count becomes low and generally negligible, except in cases of deficient operation.
- It thermally treats solids while realizing a large reduction in volume, for MSW often by a factor of 10 or more.
- It destroys gaseous and liquid waste streams leaving little or no residues, except for those linked to flue gas neutralization and treatment.
- The heat of combustion generated may be put to good use.

Disadvantages

- Incineration is technically a complex process, requiring huge investment and operating cost as well as good technical skill in maintenance and plant operation, in order to conform to modern standards.
- Heat recovery takes place under adverse conditions (boiler fouling, erosion, corrosion) and is often costly and inefficient.
- Incineration generates an amount of pollutants which are not easy to control.

This chapter was originally published as part of the Encyclopedia of Sustainability Science and Technology edited by Robert A. Meyers. doi:10.1007/978-1-4419-0851-3

A. Buekens, *Incineration Technologies*, SpringerBriefs in Applied
Sciences and Technology, DOI 10.1007/978-1-4614-5752-7_2,
© Springer Science+Business Media New York 2013

- Complete burnout of flue gas and residues needs to be ensured.
- As emission codes become more stringent, operating costs rise and the volume of secondary waste streams requiring further disposal increases (in decreasing order with dry, semi-wet, or wet gas scrubbers).

Some types of waste are banned from incinerator plants, unless they are specifically equipped to cope with such waste, e.g.:

- Volatile metal (i.e., principally mercury, thallium, and cadmium) bearing waste.
- PCB-containing waste, which requires special incinerators with unusually high destruction efficiency.
- Radioactive waste. The absence of such waste is now routinely checked in MSW, due to widespread use of medical radioactive preparations for either diagnostic or treatment purposes. Radioactive waste can be incinerated like other waste, with (a) volume reduction and (b) immobilization of radionuclides in ash as major aims; yet, containment is essential. Incineration may hence be conducted under slagging conditions. Dust filters should substantially retain all dust.

Chapter 3
Waste Incineration

Waste Incineration can be described as "the controlled burning of solid, liquid or gaseous combustible wastes so as to produce gases and residues containing little or no combustible material" (Ph. Patrick, 1980. Past president of the Waste Management Institute (UK)). The technique is now considered from various viewpoints:

- Waste streams of interest
- Phenomena in waste incineration
- Stoichiometry
- Mass balances
- Incineration products
- Residues
- Thermal aspects
- Furnace capacity
- Safety aspects
- Incinerator furnaces – principles – operations – fields of application
- Post-combustion
- Heat recovery
- Corrosion problems
- Flue gas composition and cleaning
- Dioxins

Next, the major types of incinerator furnaces and the conversion of waste into refuse-derived fuel are discussed.

This chapter was originally published as part of the Encyclopedia of Sustainability Science and Technology edited by Robert A. Meyers. doi:10.1007/978-1-4419-0851-3

A. Buekens, *Incineration Technologies*, SpringerBriefs in Applied
Sciences and Technology, DOI 10.1007/978-1-4614-5752-7_3,
© Springer Science+Business Media New York 2013

Waste Streams of Interest

Incineration generally addresses combustible waste, whether it is gaseous, liquid, sludge, paste like, melting or solid. Particular streams are municipal solid waste (MSW); commercial, industrial, and hazardous waste; sewage sludge; and hospital waste. Waste that fails being auto-combustible can still be incinerated by means of auxiliary fuel.

Municipal solid waste (MSW) has been routinely analyzed by manual sorting (and sieving of fines) in the Netherlands even on an annual basis and for different types of residential areas (TNO). Argus, Berlin, produced a very much detailed analysis in the early 1980s [32, 33]. Each major sorting fraction (fines, vegetal matter, paper and board, plastics, etc.) was analyzed for its pollutant contents (elementary composition, heavy metals, and dioxins).

Industrial process streams can be very diverse, e.g., gaseous, aqueous, and organic effluents from the most diverse industrial processes, sludge and dust from treating such effluents, waste oil and solvents, and, finally, solid waste. Process waste with stable characteristics is often disposed in-plant, in boilers, or furnaces. Occasional waste and small arising is stored in empty drums, bags, or barrels, grouped and sent to waste disposal centers. Some large factories, such as *BASF* (Ludwigshafen) or *Ford* (Cologne), have operated their own centers since the 1960s. The community operates some comprehensive centers (Denmark, Bavaria); private or public/private entrepreneurs manage others.

Green waste (branches, brush, and logs) may be collected separately for shredding and/or composting or for use in waste-to-energy (WtE) schemes.

Sewage sludge is also a generally occurring municipal waste, mainly consisting of water, so that mechanical dewatering and drying yield tremendous reduction in volume. Co-firing has been practiced many different ways, in mass burning, power plant, etc. Dedicated furnaces are mainly fluidized bed, multiple hearth, and rotary kiln.

Bulky waste or bulky refuse relates to waste types too large to be accepted by the regular waste collection, such as discarded furniture, large household appliances, and plumbing fixtures. The tendency to incinerate such items directly has declined: bulky waste is diverted increasingly for reuse and recycling; what remains is shredded before incineration. Some plants for bulky loads were operated on a full-day burning, nighttime cooling cycle. For fuel economy and especially for environmental reasons, such practices are no longer recommended. Dismantling for recycling and shredding of nonrecyclables is a better option.

Automotive shredder residue (ASR) often contains hazardous substances such as lead, cadmium, and PCBs. Some countries have classified ASR as hazardous waste and have established legislative controls.

Hospital waste is another stream often earmarked for incineration. Its composition varies with local systems for segregated collection. Specific compounds are sharps and disposables and infectious waste. Hospital waste is often incinerated in a two-step process, first partial oxidation then high-temperature post-combustion of fumes, derived from the pyrolyzer. There is a tendency to concentrate incineration in centralized units rather than in scattered and ill-managed small local plants.

Hazardous waste as a rule loses its hazardous properties during incineration. The hazards are more relevant during collection, storage, and pretreatment than during incineration proper. One category of waste stands out: *chlorinated waste* can best be fired to eliminate its persistent, lypophilic, and bioaccumulating properties. Alternatives, such as dehalogenation exist, yet are an order of magnitude more expensive. Particular aspects of chlorinated waste incineration are:

- Very high combustion efficiency ($\eta_{Comb} > 0.9999...$) is required.
- Hence, a minimum combustion temperature of 1,200°C is stipulated.
- The Deacon reaction (forming chlorine gas) is avoided by operating with minimal excess of air, possibly addition of steam (both to steer the equilibrium), and fast cooling or even water quenching (to freeze the high temperature composition).
- The formation of phosgene ($COCl_2$) is also avoided, by reducing Cl_2 formation and striving for complete conversion of CO into CO_2.

Dedicated thermal units are developed for *recovery* and cleaning of metal or metal parts, contaminated with paint, lacquers, or polymers.

Important Properties of Waste

Important properties of waste are related to:

- Storage behavior and potential hazards during storage (cfr. Safety Aspects)
- Form and size of individual particles and their distribution, physical and bulk density, specific surface, angle of repose
- Flammability and putrescence of *solid wastes*
- Bulk and physical density, viscosity, heat conductivity, reactivity, explosion limits, flash point, ignition temperature, vapor pressure, boiling point, gas evolution or decomposition during preheating, corrosiveness, toxicity, possibility of auto-oxidation, spontaneous polymerization or other incontrollable, exothermic or dangerous reactions of *liquid wastes*
- Density, explosion limits, toxicity, and corrosiveness of *gaseous wastes*.

Waste Gases – Liquids - Solids

The *heat of combustion* of pure chemical compounds is simply derived as the difference between the heat of formation of products and reactants.

The *higher heating value* (HHV) of fuel is derived by burning a known amount with oxygen in a bomb calorimeter and monitoring the amount of heat liberated that is largely transferred to the water mass surrounding the combustion chamber. The resulting temperature rise is proportional to the heat liberated; heat losses to the surroundings are corrected for by calibration. Several empirical formulas were developed to estimate the heat of combustion of a fuel, from its elementary composition, e.g., the Dulong equation (originally developed for coal and later modified to accommodate a variety of fuels, including municipal solid waste).

The heat of combustion of the combustible fraction of refuse is given by:

$$327.81(C) \ + \ 1504.1(H - O/8) + 92.59S$$
$$+ \ 49.69O + 24.36\,N \quad kJkg^{-1} \tag{3.1}$$

In this formula C, H, O, S, and N stand for the mass percent of each of these elements. These are expressed on a moisture and ash-free (maf) basis. More formulas are cited in Niessen [35].

The *lower heating value* (LHV), also termed *calorific value*, is lower, because from the HHV value one must subtract the *latent heat of condensation* of water vapor present in the flue gas, but which generally is lost with the flue gas in the plume.

The *proximate analysis* establishes the moisture and ash content (wt.%) and – by difference – the combustible part of the waste (wt.%). Thus the proximate analysis defines the amount of moisture to be evaporated prior to combustion and the required dimension of the ash handling equipment. Moist wastes, such as garbage, sewage sludge, and aqueous solutions, burn only after evaporation of most of the moisture contained. Hence, adequate measures should be taken to ensure fast and complete drying.

The *elementary* or *chemical analysis of the combustibles* should be known in order to estimate the composition of the flue gas at a given excess of air and to determine whether wet or other scrubbing of the flue gas is required.

The other properties are helpful to select and specify the waste storage, handling and feeding facilities and the required safety provisions. Information is also required on the frequency and timing of the deliveries, the kind of containers and packaging, etc.

Individual gaseous combustible compounds are characterized by means of their chemical formula and structure and molecular mass (often termed molecular weight). Density is proportional to molecular mass, which is easily derived as the sum of all atomic masses. Denser gas requires proportionally more combustion air and hence a larger supply of air to the burner. The HHV is roughly proportional to the mass of fired gas. Gases are also often characterized by their Wobbe-index, a factor combining HHV and density.

Important properties for *liquid fuels* or waste are viscosity, density, flash point, surface tension, sooting tendency, etc. These affect oil atomization and combustion, as well as burner construction, operation, and maintenance (Table 3.1).

Table 3.1 Some models for combustion of liquid and solid particles

| Model | Hypotheses | | | | |
	Number of compounds	Surface temperature compared to gas temperature	Heat exchange	Oil thermal conductivity	Diffusion in droplet
The d² law	One	Lower	Radiation	High	–
Scale model	One	Comparable		Nil	–
Homogeneous temperature	One	Comparable	Radiation + losses	High	–
Diffusion control in droplets	Several	From enthalpy balances	Rate laws	Rate laws	Species balances
Direct simulation	Several	From enthalpy balances	Rate laws	Rate laws	Species balances

Source: After Görner K [34]

Solid fuels or waste vary in chemical composition and thermal behavior. Coal consists of highly condensed aromatic structures capable of thermal softening, melting, and decomposing. Depending on its rank, coal generates combustible gas and volatiles during combustion, giving rise to flaming combustion and leaving a carbonizing residue. Biomass predominantly consists of cellulose structures, bounded by lignin. Worldwide, it is still an important fuel; yet, it loses much of its importance in terms of industrial use and trade.

Phenomena in Waste Incineration

Combustion science has evolved enormously, with respect to both theoretical concepts and experimental study. Some relevant references as well as past and ongoing conferences are cited in the general bibliography. Incineration is much more an empirical engineering science [35–38]. The last reference provides a state-of-the art review, composed on the basis of European experience.

Combustion of *flammable gas* follows two distinct modes: fast combustion in premixed flames (*mixing is burning*) and diffusion-controlled flames, those relevant in this context. Since waste flammable gases are difficult to store in oil refineries or petrochemical plants, they are commonly disposed of by either elevated or ground *flares*. Severe sooting may occur during an emergency, when large flows need to be flared. Sooting is reduced by addition of steam through ejectors located in nozzles that draws in ambient air. Smaller, better controllable gas streams are often burned in available boilers or furnaces. Where necessary, they are combusted either thermally in a dedicated yet simple combustion chamber, or catalytically on a fixed catalytic bed.

Combustible liquid wastes are generally fired, dispersed into fine droplets, each of which burns as a small entity, composed of evaporating liquid and diffusion flames around the periphery.

Solid fuels first dry, and then thermally decompose while heating, with evolving volatiles sustaining flaming combustion and the charring residue much slower glowing combustion. Converting fuel or waste into volatiles and fixed carbon is an essential step (pyrolysis) in their combustion. Mimicking this process is an essential test; for coal this test was standardized differently in each industrial country, yet 950°C is a typical temperature for defining the split between volatiles and fixed carbon. Heating rate applied and test duration also influence this split (Fig. 3.1).

In practice, these steps proceed partly in parallel, rather than in a strict sequence.

Drying is a gradual process: moisture can be absorbed quite loosely, e.g., by plastics, or firmly, physicochemically bound to its substrate.

All organic materials decompose upon heating, generating generally smaller and simpler molecules. The emerging volatiles contain inorganic (CO, CO_2, H_2O, H_2, etc.) as well as aliphatic and aromatic organic compounds; their product

Fig. 3.1 Flaming combustion of solids [39] (By courtesy of Wikipedia)

distribution depends on numerous factors, such as raw materials, temperature, residence time of volatiles and solid fraction and – not in the least – catalytic effects exerted by ash, bed material, or furnace walls.

Primary pyrolysis products show structures close to those of the molecules pyrolyzed. The longer the residence times, the more these structures evolve toward thermally more stable molecules. Ultimately, mainly carbon, hydrogen, and water vapor remain when pyrolysis is concluded in the absence of air.

Cellulosic compounds, such as paper or wood, decompose already at ca. 250°C according to quite complex mechanisms that thermally soon become self-sustaining. Some plastics, conversely, follow simple-looking *unzipping* mechanisms, yielding monomer or oligomer (low polymers, such as di-mer, tri-mer, etc.) as a product. This is the case for, e.g., polymethylmetacrylate (PMMA) and polystyrene (PS).

Vinyl compounds (polyvinylchloride, polyvinyl alcohol, polyvinyl acetate) decompose at unusually low temperatures, releasing hydrochloric acid HCl, water, and acetic acid (CH_3COOH), respectively.

PVC also decomposes in two steps. HCl evolves almost quantitatively from PVC between 225°C and 275°C. This step also produces some benzene. The second step yields further, mainly aromatic compounds, by internal cyclization [40].

Polyolefins, such as polyethylene and polypropylene pyrolysis attains a maximum rate of decomposition at ca. 450°C [41]. Primary products are polyolefinic and paraffinic chain fragments, following a Gaussian molecular weight distribution: higher temperature generates in average shorter product molecules. Secondary products from polyethylene, as well as primary products from polypropylene, show more branched chain products.

Solid waste incinerators generally feature a *mechanical grate* that supports, conveys, and pokes the waste, while primary combustion air activates the fire and cools the grate. Traveling grates, roller grates, and reciprocating grates show *plug flow* characteristics, i.e., an almost even residence time for the different refuse parcels that move through the furnace. This leads to successive zones of drying, heating, ignition, and flaming combustion of waste, and residue burnout. *Reverse-reciprocating grates* create back-mixing, by pushing the burning waste upstream, underneath the incoming fresh refuse.

Incinerators burn highly flammable plastics, side by side with wet vegetal waste. Once heated high enough (>400°C) for fast pyrolysis to occur, plastics decompose swiftly and hence burn rapidly, creating oxygen-deficient flames and leaving craters in the original refuse layer. On the other hand, wet waste is slow to ignite, for first it must be superficially dried before it can start rising in temperature, generating flammable vapors, and eventually catching fire. Even then, large lumps of moist vegetal matter may remain wet internally and survive incineration. Also, massive wood, or a thick book, takes time to burn, the carbonized material thermally insulating the flammable core.

Thus, burning refuse is heterogeneous and produces strands of oxygen-deficient hot gas as well as other gases, still prior to ignition and composed of moist air, charged with smelly products, arising in drying and heating. Unless hot oxygen-deficient and cold oxygen-rich flue gas strands are thoroughly mixed by blowing in secondary air at high speed, products of incomplete combustion will likely leave the furnace unconverted (Fig. 3.2).

Draft is the most important physical factor determining incinerator capacity. Only the smallest units operate on natural draft, as generated by the chimney. Fans (forced draft) blow in primary and secondary air; a much larger fan in front of the stack provides induced draft. The stepwise extension of flue gas cleaning, necessitated by past progression of the cleaning levels, has inflated the head losses and increased the required capacity and the power consumption of induced draft fans.

The *residence time* of gaseous and liquid wastes in an incinerator amounts to only few seconds. The residence time required for complete combustion of solids is generally about half an hour.

Hence, incinerator feed should always be made as *homogeneous* and *constant* as possible, e.g., by mixing, blending, and for municipal solid waste (MSW) ageing, to provoke moisture transfer and to account for a wide difference of flammability between easily igniting plastics on the one hand and moist vegetal waste on the other.

Fig. 3.2 Representation of a mechanical grate incinerator (By courtesy of Keppel-Seghers)

Stoichiometry

Gaseous and liquid waste can be completely combusted using *low excess of air* (5–15%) as far as their composition is sufficiently predictable and constant and mixing of air and fuel well organized. In principle, much more excess is required when firing solid waste, except in incinerator types featuring first-rate air/solids contact, e.g., fluidized bed or vortex units. Lower airflow also has other advantages: it elevates the combustion temperature, extends the residence time in a given furnace volume, and reduces entrainment of fly ash with flue gases, as well as thermal losses with flue gas in the stack.

The amount of *combustion air* provided markedly exceeds stoichiometric requirements, following from formal reaction formulas, such as:

$$C + O_2 \Rightarrow CO_2 (12g + 32g = 44g) \tag{3.2}$$

$$2H + \frac{1}{2}O_2 \Rightarrow H_2O(2g + 16g = 18g) \tag{3.3}$$

$$S + O_2 \Rightarrow SO_2(32g + 32g = 64g) \tag{3.4}$$

Combustion equations are normally marked in atomic or mol units; the corresponding weight amounts are marked in grams. Under standard conditions 1 mol of gas has a volume of 22.4 l or dm^3.

Mass Balances

The *Law of Mass Conservation* also applies to incineration. Hence, the sum of all *input streams* equals the sum of all *output streams*, whether

- In *total* mass flows (kg/h)
- *Any individual element* entering and leaving the plant, and expressed either in mass units (kg/h) or in number of moles (mol/h). In combustible waste, the main elements (symbol, atomic mass) are carbon (C, 12), hydrogen (H, 1), oxygen (O, 16), sulfur (S, 32), nitrogen (N, 14), and chlorine (Cl, 35.5).

Input streams are typically (1) waste, (2) auxiliary fuel (when needed), and (3) primary and secondary combustion air and also uncontrolled air entering through leaks. The latter can be estimated along the flue gas path, simply by measuring the rising oxygen or the declining carbon dioxide content of the flue gas.

During *flue gas cleaning*, additional compounds may be added, such as basic additives (hydrated lime $Ca(OH)_2$, lime CaO, or even – at high temperature – ground limestone $CaCO_3$, and also sodium bicarbonate $NaHCO_3$ or hydroxide NaOH), ammonia or urea (DeNOx), and activated carbon, as an adsorbent for principal organic pollutants (cfr. Flue Gas Treatment).

Typical *output streams* are grate siftings, bottom ash, boiler slag, fly ash, flue gas neutralization residues, and cleaned flue gas. In some plants, the different flue gas treatment residues are extracted as a mixture, in others separately.

Mass balances directly relate input streams to output streams.

One Mg (tonne) of MSW requires 6.5–7.8 Mg (5,000–6,000 Nm^3) of combustion air. Typically, mechanical grate incineration generates (EU conditions):

- 250–350 kg of bottom ash
- 5–15 kg of boiler slag
- 20–40 kg of fly ash
- 5–15 kg of neutralization salts
- 7–8.6 Mg of flue gas

Fig. 3.3 The Tanner diagram [37]

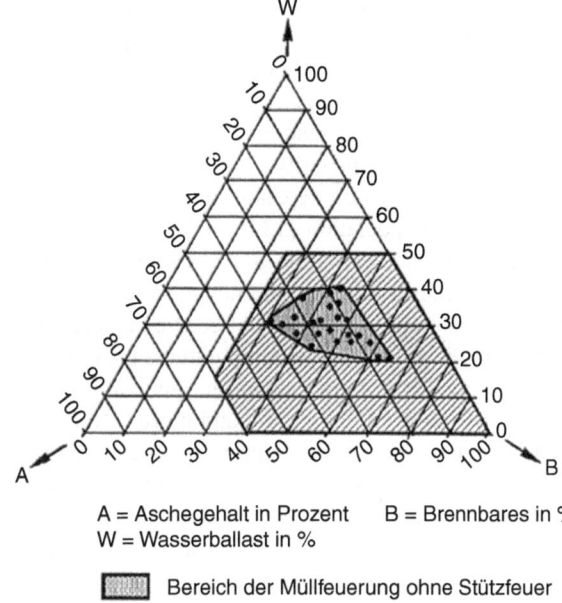

A = Aschegehalt in Prozent   B = Brennbares in %
W = Wasserballast in %

Bereich der Müllfeuerung ohne Stützfeuer

Incineration Products

On the basis of aforementioned balances, the amount and identity of the incineration products can now be derived.

The *proximate analysis* splits the waste to be fired up into three parts: (1) moisture W, (2) ash A, and (3) combustibles C. The first reports to the flue gas, the second forms the residues, whereas the third is converted into combustion products also reporting to the flue gas. Starting from W% + A% + C% = 100, Tanner represented waste composition in a triangular diagram, in which a zone of auto-combustible MSW is identified (Fig. 3.3).

Slight disparity occurs between ashes, as originally present in waste, the "real" ash resulting during the proximate analysis test and that formed during incineration proper. Depending on the ash minerals on the one hand and the combustion conditions on the other, the original ash may differ from actual incinerator ash, because of occurrence of various thermal reactions, such as,

$$Ca(OH)_2 \Rightarrow CaO + H_2O \tag{3.5}$$

$$CaCO_3 \Rightarrow CaO + CO_2 \tag{3.6}$$

$$CaSO_4 + C \Rightarrow CaO + SO_2 + \frac{1}{2}CO \tag{3.7}$$

as well as many others that can only be identified by a detailed study of the ash minerals through methods such as X-ray fluorescence (XRF) or scanning electron microscopy (SEM), their thermodynamic stability, potential reactions, and state of conversion. Generic classes of such reactions are: dehydration, decarbonatation, sulfate decomposition, and decomposition of higher oxides into lower oxides. Another reason for disparity is the occurrence of volatilization at flame temperature; such volatilization depends on temperature, presence of oxidizing or reducing conditions and speciation [124]. Halogenides (chloride, bromide) are much more volatile than oxides or sulfides, carbonates, and sulfates.

Coarse or sintered ash materials report to the bottom ash, fines are at risk to be entrained. Bottom ash consists of coarse objects, such as stones, glass, or cans, and of ash proper. Low burnout temperatures preserve the original ash structures; high temperatures first cause sintering, generating larger and more solid sintered structures, and eventually more and more fusion.

The major gaseous incineration products are carbon dioxide and, to a variable extent, water vapor and, of course, a large amount of air nitrogen. *Carbon dioxide* generation is directly proportional to the amount of carbon burned, since the background carbon dioxide content in combustion air is insignificant (0.03 vol.%), compared to carbon dioxide in flue gas. This carbon dioxide concentration varies widely, from few vol.% to about 12 vol.%, depending on waste composition and on the excess air used.

Incomplete combustion leads to the formation of *carbon monoxide (CO)*, *total organic carbon (TOC)*, and *black carbon (BC)* or *soot*. The amount of carbon monoxide formed is highly variable, with generally a stable background value, spiked by rare or more frequent peaks (from less than 1 ppm to peaks of some 10,000 ppm, or 1 vol.%, occurring only during combustion upsets), yet only rarely influences the carbon dioxide content (Fig. 3.4).

The *moisture content* of flue gas is composed of:

- The original *moisture content* of the fuel fired, which upon drying reports to the flue gas. This is normally negligible for oil and gas fuels, but it reaches several percentages for powdered coal, and is quite substantial for peat, lignite, most forms of biomass, such as sewage sludge or green wood, and for municipal solid waste (MSW).
- *Moisture* contained in *combustion air*, varying markedly with both temperature and relative humidity.
- *Chemically formed water*, derived from the hydrogen content of fuel. The amount can easily be derived by simple stoichiometric computations, based on reaction equations such as:

$$[C_6H_{10}O_5]_n + 6nO_2 \Rightarrow 6nCO_2 + 5nH_2O \qquad (3.8)$$

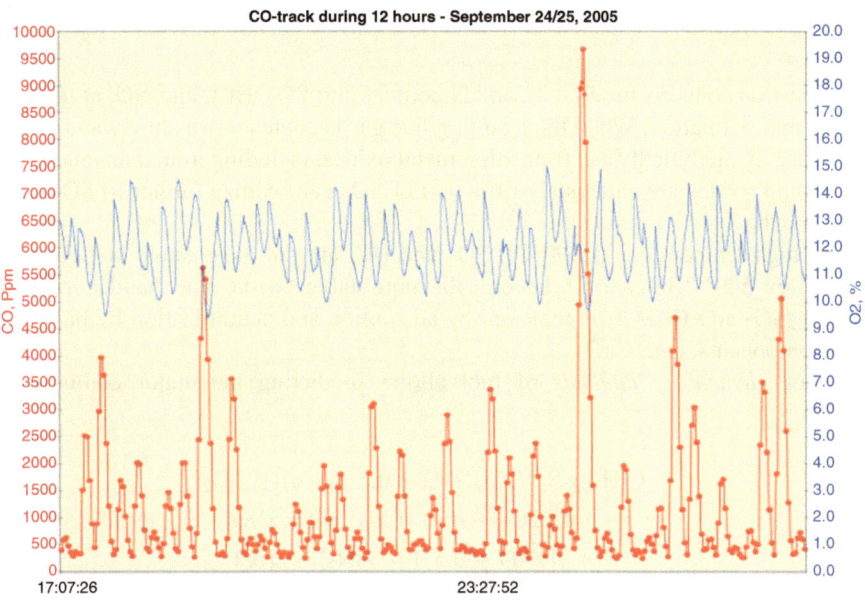

Fig. 3.4 Time evolution of carbon monoxide as a function of time [42]

with, e.g., five volumes of water vapor formed per anhydro-cellulose unit $C_6H_{10}O_5$ (the cellulose monomer) fired, or in mass units 90 g of water vapor formed per 162 g of solid fuel.

- Water added and evaporated during *quenching* of flue gas by water injection, a usual practice in small incinerators and in the incineration of chlorinated waste.
- *Pre-conditioning* of flue gas, prior to scrubbing, to enhance the elimination of fine dust, HCl, and SO_2. The first become denser, the acid gases are absorbed more easily by hydrated lime in the presence of water vapor.
- Water evaporated in wet scrubbers, used for scrubbing out acidic gases. This treatment saturates the flue gas with water vapor; the resulting temperature is typically 65°C. Sometimes, the scrubbing water is cooled by heat exchangers to obtain a deeper separation of various pollutants (e.g., mercury, soluble gases, and organic vapors).

Oxygen, present in the fuel, reduces the amount of combustion air required, but does not contribute to the heating value. Heteroatoms, such as sulfur, nitrogen, chlorine, and other halogens may contribute to air pollution, since they are converted largely (sulfur, chlorine) or partly (nitrogen) into pollutants. Still, flue gas cleaning will eliminate the resulting pollutants, down to the limit values specified (cfr. Tables 3.2 and 3.3).

Formation of Pollutants

Combustion converts the S-, Cl-, and N-content into SO_2, HCl, and NO, at least as a first approximation. When the resulting flue gas is cooled down slowly and in the presence of catalytic fly ash (transition metal oxides, including iron, manganese, or vanadium oxides are catalysts), a fraction of SO_2 can oxidize further to SO_3, and HCl to Cl_2.

At high temperature ($1{,}000°C$), SO_2 and HCl are the most stable compounds; yet, below $500°C$ SO_3 and Cl_2 become the more stable. On the other hand, a fraction of SO_2/SO_3 and HCl/Cl_2 is removed by adsorption and neutralization by basic fly ash components, e.g., CaO.

Thus *elementary analysis* of fuel allows predicting the major combustion products:

$$C_aH_bO_cS_xN_yCl_z + [a + 0.5(b - c) + x]O_2$$
$$\Rightarrow aCO_2 + (b - 0.5z)H_2O + xSO_2$$
$$+ y[\alpha NO + 0.5(1 - \alpha)N_2] + zHCl \qquad (3.9)$$

Nitrogen oxide (NO) forms from *fuel*-N (i.e., the organic nitrogen, e.g., from proteins, in sewage sludge, hair or leather, or from polyamides) and also from combustion air, yet mainly at elevated temperatures, as *thermal* NO. Such NO formation is lower when the flame is cooled, e.g., by radiant heat losses or by the presence of water vapor, and also when combustion is conducted in two steps: the first under reducing conditions, the second oxidizing, yet at low temperature.

To cater for this uncertainty, fuel NO formation is given a proportion α ($0 < \alpha < 1$), the balance being reduced or decomposed to molecular nitrogen ($1 - \alpha$). The formation of thermal NO is neglected in Eq. 3.9 (Fig. 3.5).

Chlorinated Compounds

Most waste contains chlorides and also chlorinated organic compounds.

The Bundesweite Hausmüllanalyse (comprehensive analysis of refuse and its sorting fractions in the German Federal Republic) established the amount of, e.g., heavy metals, PAH, and dioxins in MSW for fractions such as fines, vegetal, synthetic, paper, and board. All these sorting fractions are contaminated with all kinds of pollutants [32, 33].

During incineration, organic compounds are destroyed and their chlorine content is converted to HCl. Typically 50% of the Cl-content comes from PVC [38]. During combustion, PVC, as well as a vast majority of organic and inorganic chlorinated compounds, is partly or completely converted into HCl. PVC liberates HCl very easily. Such release is also likely to be complete, unless some other compounds,

Fig. 3.5 NO as a function of combustion temperature [37]

EOLSS - POLLUTION CONTROL THROUGH EFFICIENT COMBUSTION TECHNOLOGY

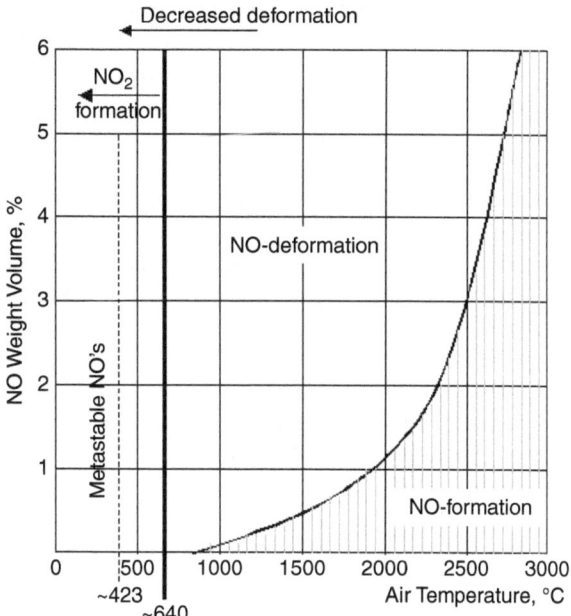

e.g., $CaCO_3$, capture it; the latter is plausible in numerous applications featuring fillers of precipitated calcium carbonate or ground dolomite/calcite.

The presence of NaCl is ubiquitous, especially in marine surroundings. At high temperatures, NaCl reacts with steam, yet its conversion into NaOH and HCl is limited by thermodynamic equilibrium. It shifts largely to the right, however, in case NaOH is itself converted into silicates, aluminates, or other composite compounds [43, 44], e.g.:

$$NaCl + H_2O \Rightarrow NaOH + HCl \tag{3.10}$$

$$2x NaOH + y\, SiO_2 \Rightarrow x Na_2O\, y\, SiO_2 + H_2O \tag{3.11}$$

$$2x NaOH + z\, Al_2O_3 \Rightarrow x Na_2O\, z\, Al_2O_3 + H_2O \tag{3.12}$$

Thermodynamically, the formation of chlorine gas from hydrogen chloride is described by the industrially important Deacon equilibrium:

$$4HCl + O_2 = 2Cl_2 + 2H_2O + Heat \tag{3.13}$$

At combustion temperatures, HCl is by far the main Cl-compound yet – below 500°C – equilibrium conditions reverse and elemental chlorine gains ground. Chlorine is much more reactive and corrosive; moreover, it is slower to dissolve in water and thus difficult to remove. Fortunately, this reaction also becomes slower and slower, so that there is little progress towards equilibrium during the few seconds while flue gas moves from furnace to stack. The Deacon reaction also shows an effect of oxygen partial pressure, an even stronger effect of water vapor, as well as an effect of total pressure.

Other halogens follow similar equilibriums, with the elementary amount rising in a sequence: $F_2 < Cl_2 < Br_2 < I_2$. The Deacon reaction is a potential source of both corrosion and dioxin. No doubt, chlorine is only rarely produced in significant quantities and only in the presence of oxidants, such as iron ore (Fe_2O_3) or manganese ore (MnO_2).

In industry, the Deacon process is of paramount importance: chlorine is a potent reactant required in organic chemistry and synthesis. Its use leaves HCl as a useless by-product. However, reaction (Eq. 3.13) allows recovering chlorine from HCl. Typical reaction conditions are: fixed or fluid bed, 450°C, $CuCl_2$ catalyst, and in dry air or pure oxygen.

Residues

In principle, incinerator residues are inert and sterile. Often, the major components in ash are silica (SiO_2), alumina (Al_2O_3), and lime (CaO), which are also the main components of the earth crust; yet virtually all elements are represented and ash composition may differ greatly from that of the earth, especially in industrial waste. Numerous studies have been devoted to the physical nature and the minerals of bottom ash and fly ash [24–26, 45, 46]. The International Ash Working Group merged worldwide experience in characterization, treatment, and leaching tests for evaluation of eventual environmental impacts of incinerator residues. Fly ash, a by-product of fossil fuel firing (coal, lignite, peat) is the subject of a site of Kentucky University and of periodic conferences published there.

Chemical analysis of the mineral ash gives information on the softening and melting behavior of the ash and hence about its tackiness and possible attack on refractory. As a rule, Na- and K-compounds decrease the melting point, in particular when present as persulfates, vanadates, borates, etc. The same holds for fluxing elements, such as boron, vanadium, or fluor. The presence of volatile compounds, such as mercury, thallium, cadmium, arsenic, antimony, and other volatile heavy metals makes the related wastes improper for incineration in conventional units. In numerous cases, stable mineral forms are different at the conditions of high-temperature combustion and at room temperature, e.g., volatile chlorides, stable at combustion temperature, tend to convert into sulfates once they condense on boiler tubes.

Thermal Aspects

During incineration, the heat content of waste, in particular its higher heating value (HHV), is liberated almost entirely. The only exceptions are the unburned materials in bottom ash, fly ash, and flue gas. Combustion efficiency η_{Comb} addresses these chemical losses by:

$$\eta_{Comb} = 1 - Ash\,C_{ash}HHV_C - (Fly\ Ash)C_{fly\ ash}HHV_C \\ - (TOC)HHV_{TOC} \tag{3.14}$$

An incinerator plant is a thermal plant and should be operated as evenly and constantly as possible, close to the setpoint in its operating diagram (Fig. 3.6). Capacity is expressed either as (nominal) thermal load (GJ/h), or as weight throughput (Mg/h).

The *operating temperature* of an incinerator combustion chamber can be estimated from a heat balance and depends on:

- The higher heating value of waste
- The *excess air* applied

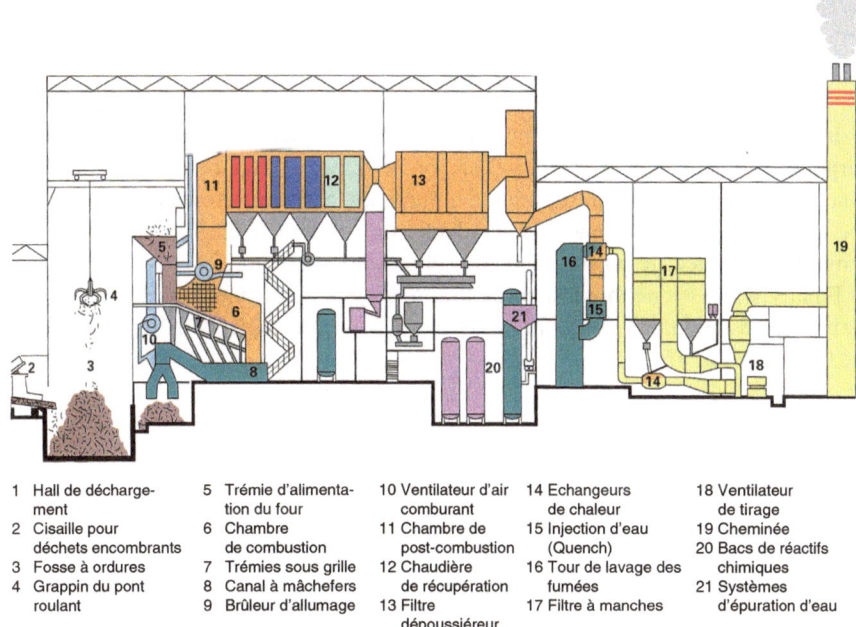

1 Hall de décharge-ment	5 Trémie d'alimenta-tion du four	10 Ventilateur d'air comburant	14 Echangeurs de chaleur	18 Ventilateur de tirage
2 Cisaille pour déchets encombrants	6 Chambre de combustion	11 Chambre de post-combustion	15 Injection d'eau (Quench)	19 Cheminée
3 Fosse à ordures	7 Trémies sous grille	12 Chaudière de récupération	16 Tour de lavage des fumées	20 Bacs de réactifs chimiques
4 Grappin du pont roulant	8 Canal à mâchefers	13 Filtre dépoussiéreur	17 Filtre à manches	21 Systèmes d'épuration d'eau
	9 Brûleur d'allumage			

Fig. 3.6 Operating diagram of a mechanical grate incinerator (After [37])

- The cooling of furnace walls (e.g., by tubes of an integrated boiler or by heat losses to the environment)
- The initial temperature of air and waste streams

A theoretical *flame temperature* (°C) can be derived simply by plotting the heat content of flue gas (MJ/Nm^3 × Nm^3/kg waste) as a function of temperature: the flue gas reaches the theoretical flame temperature when its sensible heat equals the liberated heat of combustion (MJ/kg waste). A more complete *heat balance* over the furnace, the boiler, and all downstream equipment gives:

$$Q_{fuel} + H_{fuel} + H_{air} = (H.H.V.)F_{fuel}$$
$$= Q_{heat\ duties} + Q_{wall\ losses} + Q_{sensible\ heat} \qquad (3.15)$$

The first three terms contain the chemical energy (Q_{fuel}) liberated by combustion, augmented by the enthalpies of fuel (H_{fuel}) and air (H_{air}) when entering the furnace. After combustion, the energy entering the furnace is eventually redistributed as:

- *Useful energy* ($Q_{heat\ duties}$), taken up by the various heat duties, generally the boiler, the economizer, and the air preheater
- *Wall losses* ($Q_{wall\ losses}$) by convection and radiation
- *Sensible heat* ($Q_{sensible\ heat}$) and *latent heat* (water vapor) contained in the flue gas at the stack, i.e., the stack losses

Thermal efficiency η_{Therm} addresses these wall losses and stack losses by:

$$\eta_{Therm} = 1 - Q_{wall\ losses} + Q_{sensible\ heat} \qquad (3.16)$$

It indicates the fraction of the energy entering that is recovered for useful purposes. Typical values are 0.6–0.85, or 60–85%. It can be used for district heating, water desalination, or industrial purposes. Since all these applications are site dependent and not generally available, the heat recovered as steam can be converted into electric power, by means of a turbo-alternator.

Finally, there is one more important ratio, indicating the yield of electric power derived from the initial energy in MSW or other waste incinerated. Typical values are 0.16–0.24, or 16–24%.

Air Preheating

Primary air preheating facilitates ignition, increases the flame and combustion temperature, and improves the thermal balance of the process by recovering more heat from flue gas. Combustion air is often preheated, either by flue gas/air or by steam/air heat exchangers, to assist in drying and ignition. Such heat exchangers are relatively voluminous (gas/gas heat transfer is slow) and hence capital intensive.

Combustion air may also be replaced by oxygen-enriched air, or even by pure oxygen, in order to improve and accelerate combustion. This is an unusual option, since combustion air is free of charge and pure oxygen is expensive.

Low operating temperatures lead to slower and less complete combustion; excessive temperatures may render combustion control more difficult and cause severe slagging of ash and fly ash. Tacky ash gradually builds up onto the furnace walls, the deposits eventually limiting the throughput of the furnace. Similarly, clogging problems may occur in the convection sections of the boiler, when excessive approach velocities are practiced or insufficient tube clearance is provided.

Some plants operate under *slagging conditions*, at temperatures at which the ash is molten and tapped in that state. It is important to ensure steady slag flow by:

- Carefully controlling the composition of the ash, at or close to a suitable eutectic composition; iron silicates and glass are two examples of compositions with accessible melting point
- Providing auxiliary burners and, when required, adding fluxes such as fluorspar, to enhance slag fluidity

Furnace Capacity

Nominal capacity is often expressed as the *throughput* or weight capacity (Mg/h) at which the incinerator was designed. The *load factor* of the incinerator is the ratio of the actual operating rate (Mg/h) to the nominal one (Mg/h).

Incinerator furnaces are characterized best by a minimum and a maximum *thermal capacity* (MW). Below its minimum value, the heat generation rate is so low that the furnace no longer reaches adequate temperatures to ensure smooth drying, heating, and ignition, and eventually complete combustion. When the flue gas temperatures descend below 850°C, European Union Codes stipulate that auxiliary burners must ignite and heat the combusting gases, to ensure their sufficient burnout. Excessive combustion temperatures are also undesirable, because fly ash becomes too tacky, creating deposits on furnace walls and boiler tubes. Ash similarly starts slagging; the resulting deposits on the furnace walls become ampler and ampler, eventually even restricting the movement of waste on a grate.

Furnaces feature also a minimum weight capacity (Mg/h), dictated by the necessity to maintain some minimum coverage of the grate for protecting it against furnace radiation and atmosphere. Maximum related to bed density. Finally, the relation between thermal and weight capacity is also bounded, by the necessity of producing sufficient heat for heating furnace and waste; the ratio represents the heat of combustion (MJ/kg). These different boundary conditions are represented in thermal capacity vs. weight capacity diagrams, indicating the area of smooth operation of the plant. The latter is possibly extended toward low heating values by preheating combustion air or toward high-calorific waste by cooling the combustion chamber. Thus, there are links between furnace requirements and waste characteristics.

The *volumetric heat release rate* (MJ/Nm^3, s) of a given furnace is mainly determined by the quality of contact with combustion air and by fuel reactivity, which generally decreases with larger size, higher moisture content, and lower HHV. Since combustion intensity is often unevenly distributed over the furnace, the method to consider furnace volume should be carefully defined, when citing values for volumetric heat release rates. In some cases this volume has been defined as the furnace volume at temperatures exceeding 850°C, rather than as a physical geometric volume of the combustion chamber. Dead zones at lower temperature indeed may consume a sizeable fraction of furnace volume, thus reducing real residence times and combustion efficiency η_{Comb}. Conversely, the first flue of a waste heat boiler may operate above 850°C and thus become eligible as supplemental furnace volume.

The operating domain and the limits of furnace operation may be dictated by various considerations, e.g.:

- *Heat balances*, and the concomitant higher and lower *temperature* limit (°C)
- Excessive, adequate, or insufficient *thermal load* (GJ/h)
- Adequate coverage of a mechanical grates, and hence maximum and minimum *feeding rate* (Mg/h)
- Provision of sufficient combustion air

During reception tests, the operators were supposed to deliver the proof of capacity of a given incinerator furnace over a time period of 24 h. Realizing their presumable failure, they started overcharging the furnace, bringing in more and more MSW. Due to the excessive thermal load, the furnace interior evolved from orange-red to orange, then to yellow, then turning whiter and whiter as the furnace temperature rose. Still, at that moment, more and more unburned materials appeared among the residue: remarkably, a telephone book had crossed this furnace without even starting to convert into char!

Hazardous Waste

Hazardous waste can be identified either on the basis of inclusive lists, as proposed by the European Union [47], or on the basis of hazardous properties, an approach followed by the US EPA. In the USA, hazardous waste is waste that poses substantial or potential threats to public health or the environment. There are four factors that determine whether or not a substance is hazardous [48]:

- Ignitability (i.e., flammable)
- Reactivity
- Corrosivity
- Toxicity

The US Resource Conservation and Recovery Act (RCRA) additionally describes "hazardous waste" as waste that has the potential to [48]:

- Cause, or significantly contribute to, an increase in mortality (death) or an increase in serious
- Irreversible, or incapacitating reversible, illness
- Pose a substantial (present or potential) hazard to human health or the environment when improperly treated, stored, transported, or disposed of, or otherwise managed

Most of these hazards are entirely eliminated by incineration. Hence, HW may be incinerated at high temperature. Many cement kilns burn hazardous wastes like used oils or solvents. A more detailed discussion is to be found in various books listed at the end of this entry and in [49]. Hazardous waste poses much more problems at the levels of collection, bulking up (i.e., grouping similar waste in the same container or vessel), transportation, and intermediate or final storage than at that of incineration. Obviously, flue gas cleaning must take into account the chemical composition of the hazardous waste concerned.

Safety Aspects

Swiss Re provided a systematic discussion of some safety problems and accidents in incinerator plants. At the times of construction and annual maintenance of incinerators, lots of unusual activities take place onsite, bringing various hazards with them. During normal operation, these hazards reduce to more normal proportions, yet, numerous safety problems may occur around incinerator plants; just to name a few [50]:

- Waste bunker fires
- Explosions during the shredding of waste
- Flame flashback into the system of feeding locks
- Explosive combustion by simultaneous ignition of a large mass of waste, bringing the furnace under overpressure, with flames sorting out
- Hydrogen explosions following decomposition of water in contact with hot metal in a wet ash extractor
- Pressure vessels (boiler)
- Low levels of boiler feed water
- Boiler corrosion and tube failure
- Accidents connected to chemicals on-site, e.g., boiler feedwater treatment acids and bases and ammonia for DeNOx operation
- Rotary and moving equipment
- Transformer fires
- Fires in the wet scrubber, during shutdown

An even larger array of accidents may take place in plants treating hazardous waste, as a consequence of chemical reactivity, flammability, and corrosivity. During collection and storage it is usual practice bulking up liquid waste of similar composition and origin. Mixing distinct waste streams often leads to undesirable events; to avoid such happenings it is desirable to consult compatibility charts and data, such as [51–56], and also to mix small amounts in a test tube and then observe carefully any heating, gas evolution, precipitate formation, or other processes taking place.

Pool burning, *boiling liquid expanding vapor explosions* (BLEVEs) and *vapor cloud explosions* (VCE) are relevant concepts in industrial safety techniques [57]. Even comprehensive waste treatment centers do not necessarily reach the scale of operations or storage required to resort under COMAH eligibility conditions, although specific risk derives from the multitude and variability of waste streams potentially handled. Fires at chemical storage sites are generally impressive and the storage, blending, and feed preparation facilities upfront a chemical waste incinerator are exposed to such occurrences.

Chapter 4
Incinerator Furnaces and Boilers

Furnaces, Their Duties, Peripherals, Operation, Design, and Control

Most problems with incinerator plant proper are basically mechanical and arise mainly at two levels: (a) the introduction of waste into the furnace and (b) the extraction of the various combustion residues. Both should proceed without undesirable and uncontrolled entrance of ambient air.

Duties

Basically, a furnace is a heat-resistant enclosed space that should fulfill several duties simultaneously:

- Limiting the heat losses to the surroundings (heat losses $\Rightarrow$ flame cooling $\Rightarrow$ incomplete combustion).
- Ensuring controlled entries to primary and secondary combustion air, and exclude any notable uncontrolled entries, e.g., through the feeding or the ash removal system.
- Ensuring sufficient combustion + post-combustion time to both flue gas and solid phase (fuel, ash) to allow for their thorough and controlled burnout. This implies avoidance of short-circuiting, as well as creation of dead corners.
- Providing peripheral facilities for feeding the various waste streams to be incinerated and (when required) ash removal facilities.

This chapter was originally published as part of the Encyclopedia of Sustainability Science and Technology edited by Robert A. Meyers. doi:10.1007/978-1-4419-0851-3

A. Buekens, *Incineration Technologies*, SpringerBriefs in Applied Sciences and Technology, DOI 10.1007/978-1-4614-5752-7_4, © Springer Science+Business Media New York 2013

Feeding Equipment

Fuel feeding peripherals strongly depend on fuel characteristics, such as the state of aggregation of the waste to be fired in a primary combustion chamber. Examples are a conventional or more specialized burner for firing gas, liquid, or pulverized, coal, in case of flammable waste gases, pumpable waste liquids, molten solids, and finely divided, free flowing solids. Burners for liquid waste may be based on centrifugal dispersion (rotary cup burners) or on pressure or auxiliary medium (steam, high pressure air) dispersion.

Chlorinated waste has been fired using the dispersion provided by a patented small auxiliary burner situated inside the main burner: the liquid chlorinated waste is supplied through apertures in a duct, leading the combustion productions from the auxiliary burner into the main combustion chamber (Vicarb technology). Some burners are even built to receive several types of wastes simultaneously, such as waste oil, emulsions, suspensions, as well as auxiliary fuel, to sustain combustion.

Solid waste can be fired by means of:

- Gravity feeding from a fuel hopper, separated from the furnace by means of a lock, composed of two sliding doors, a rotary valve, or even a pile of waste locking out the ambient air.
- Spreader stokers [59]
- Screw or piston feeders
- Mechanical or traveling grate stokers
- Pneumatic feeding of free-flowing fuel, e.g., to cyclonic or fluidized bed combustors

Cooling and extinguishing provisions may be required for preventing backfire in feeding systems, or excessive thermal decomposition in feed lines. Another frequent issue is the presence of oversized materials, metal pieces, etc., that create problems during feeding and/or residue extraction: waste containers seem to exert a fatal attraction to all kinds of extraneous matter that can block or even destroy the most sophisticated mechanical feeding or residue extraction equipment. Operators should scrutinize incinerator feed for items such as pressurized gas bottles, ammunition, or oversized concrete or metal parts.

Ash Extraction

Dry or wet ash extraction equipment is generally installed at the bottom of an ash pit or of a sequence of these ash pits, located below successive sections of a grate. It may be based on drag conveyors with suspended flights, screw conveyors, inclined vibrating conveyors, or even pneumatic conveying. These systems must be designed as a function of flow rate and the handling characteristics of fuel and ash. Failing feeding or extraction mechanisms can cause undesirable, expensive downtime (1 day of a commercial incinerator line typically costs US $20,000–$50,000).

Dry extraction plant is somewhat simpler to maintain, yet tends to be a source of persistent dust in and around the basement, where it is located. Dry extractors create considerable chimney effects and – as a consequence – they may turn into an unwelcome source of uncontrolled air in the furnace.

Wet extraction has the merit of quenching the residue and at the same time it brings in some water vapor at the level of the discharge point. Discharging hot metal may decompose water, forming potentially explosive hydrogen.

Air Supply

The combustion chamber provides suitable plenum chambers for *primary* and entrance ports for *secondary air*, supplied at possibly substantial overpressure. Primary air activates the fire, burns out combustion residues, and cools the mechanical grate, if existing. Secondary air is injected at a high speed (typically 80–150 m/s), providing the required momentum for thorough mixing of flue gas and completing their burnout. As capacity is scaled up, the available momentum declines relative to the dimensions of the furnace. Some furnace suppliers also bring in secondary air through hollow beams, situated at the level of the furnace outlet: the secondary air is split into four parts, some supplied though nozzles situated in the side walls, the remaining from the hollow beam in the middle of the furnace exit (Fig. 4.1).

Flow Patterns

The flow patterns in a combustion chamber are rather complex, determined by the momentum of all inputs (burners, primary and secondary air) and outputs (extraction of combusting gas), as well as by buoyancy effects caused by flames and the hot combusting gas generated.

Whatever the geometry, there is strong tendency toward short-circuiting between, on the one hand, the point(s) of entry and, on the other hand, the point(s) of exit. Short-circuiting is minimal in a perfect plug flow furnace. It becomes important in the case of a voluminous combustion chamber with single entry and single exit, strong short-circuiting between entry and exit, and inactive zones in between furnace walls and short-circuit flows (Fig. 4.2).

A short-circuiting combustion chamber is inefficient: the short-circuiting threads show a very low, reduced residence time, the short-circuited volumes unduly long residence times, albeit at low combustion rates and temperatures. Hence, both are inefficient.

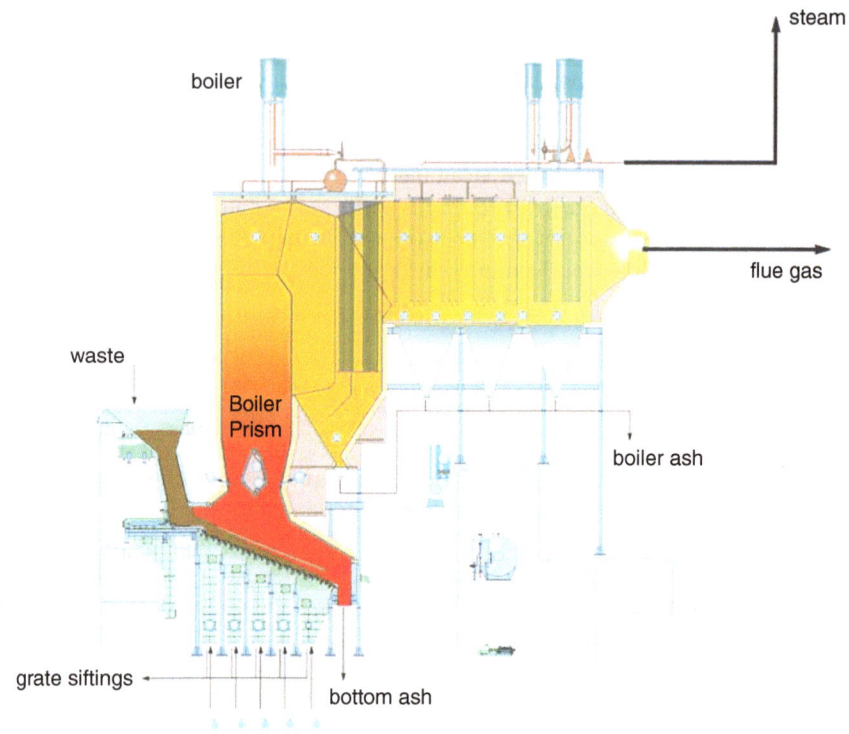

Fig. 4.1 Secondary air distribution beam in the middle of the exit from a combustion chamber (Courtesy of Keppel-Seghers, Willebroek [Belgium])

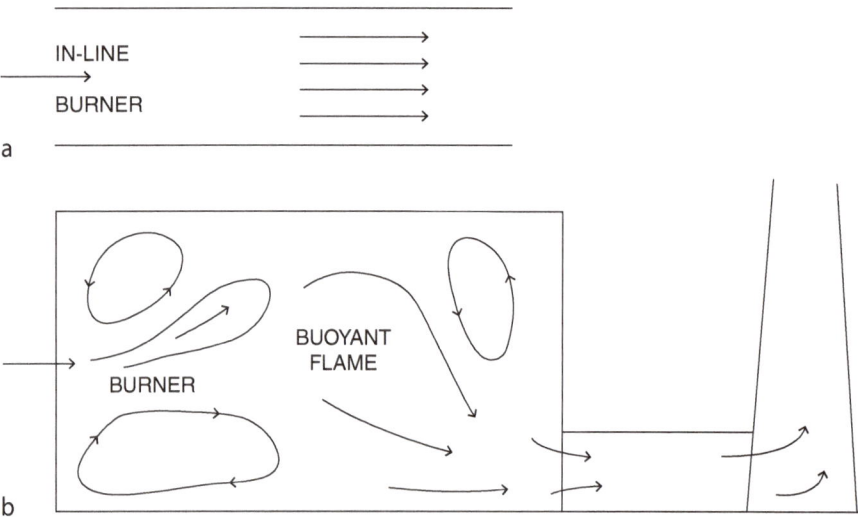

Fig. 4.2 Plug flow versus plug flow with dead zones

Flow patterns can be influenced by combustion chamber geometry, positioning of input and exit locations, selection of input momentum, and influencing the combusting gas pathways, e.g., by provision of baffles and changes in direction.

Design Aspects

Thirty years ago, only an empirical approach was practicable when designing incinerators. Tanner devised a triangle diagram to represent MSW as a ternary mixture, indicating zones with auto-combustible waste and others where auxiliary fuel was needed; Hämmerli proposed different nomograms for comparing and assessing grate loadings for mechanical stokers and rotary kilns. Today, *computer fluid dynamics* (CFD) easily derives the flow and mixing characteristics, the rates of heat generation, and the temperature and flow fields [58, 60].

Moreover, the trajectories of particles of various sizes can be predicted stochastically. Swithenbank et al. modeled the various zones (drying, pyrolysis, gasification, incineration) of a mechanical grate incinerator, using CFD, as well as the results of experimental testing at different scales [61–63]. A representative list of recent SUWIC work is given in [64]. Other important sources of solid waste incineration test work are due to ForschungsZentrum Karlsruhe, with experimental research on units such as TAMARA (small mechanical grate incinerator unit) and THERESA (rotary kiln incinerator unit).

Computer Fluid Dynamics (CFD)

Computer fluid dynamics are based on subdividing the volume of interest, i.e., the combustion chamber (or other parts of the plant) into a grid of elementary volumes. The relevant equations of conservation (mass, momentum, energy) are then applied to each of those elements, after defining all inputs, outputs, and boundary conditions. The resulting system is integrated from start to finish, after introducing momentum, mass, and heat transfer (adapted from the Laws of Newton, Fick, Fourier, and Stefan-Boltzmann), taking into account dimensional analysis, turbulent flow, and the state functions of relevant compounds, as well as chemical kinetic reaction systems of variable complexity [60].

CFD thus allows visualizing some cardinal aspects of the combustion chamber, i.e., the fluid flow field (flow vectors, indicating flow direction, and rate in each point), temperature and pressure field, and combustion rate field and – depending on nature and composition of the reaction models – fields for any other chemical compounds of interest (PICs, specific pollutants). Modeling thermal behavior of specific compounds or waste can be conducted at a milligram or even a lower scale [65, 66] (Fig. 4.3).

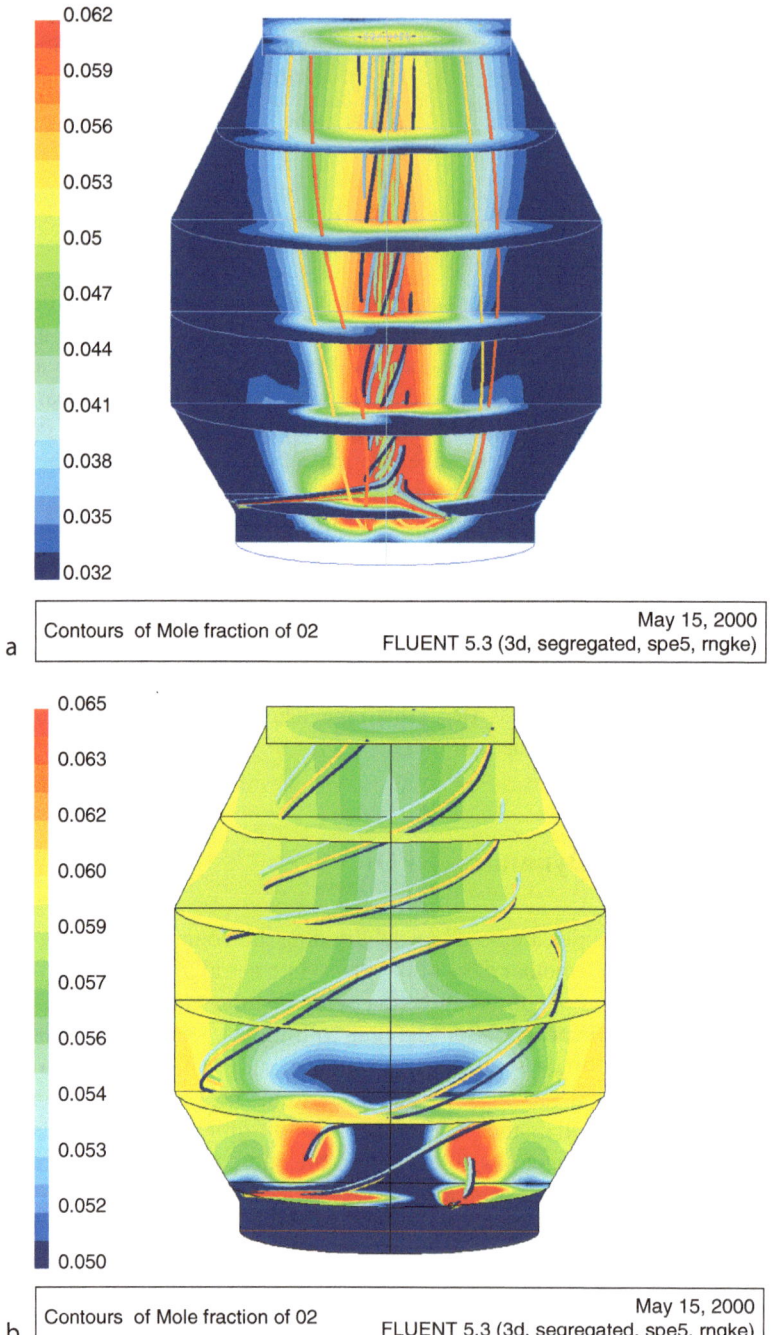

Fig. 4.3 Computer fluid dynamics (CFD) representation of a combustion chamber (By courtesy of Prof. J. Swithenbank [SUWIC])

Draft Considerations

An incinerator plant usually operates under balanced draft: a balance is struck between forced draft (blowing in combustion air) and induced draft (ID, drawing out flue gas through the stack). ID arises by means of chimney draft, supplemented by the ID-fan, so that the furnace operates steadily with a combustion chamber at a slight subatmospheric pressure, of the order of say 10 or 15 cm water column (1 atm equals more than 10 m w.c.).

Chimney draft follows from the *Law of Archimedes*: the stack is filled with light hot gas, taking the place of an equivalent physical volume of much denser ambient air. Hence, the hot stack gas aspires being replaced by the latter, which enters the furnace by all controlled inlet ports, as well as by those uncontrolled, such as a dry ash extractor or non-tight junctions between distinct parts of the plant and non-tight plant parts, e.g., a fly ash discharge valve.

Very small plants (such as a big stove) may rely on *natural draft*, controlled by means of variable obstructions regulating at the supply side or at the chimney. Medium and large plants use both *forced* and *induced draft fans*. These are major consumers of electric power. Due to the gradual extension of heat recovery and pollution control, these draft requirements have steadily risen over time. For example, power consumption in mechanical grate plant was typically 40–80 kWh/Mg of MSW around 1970. Today, it is more like 160–240 kWh/Mg of MSW.

Mechanical Drives

Until the 1920s, loading the furnace, poking the fire, and extracting ashes was largely manual, somewhat aided by gravity and appropriate tools. Mechanical grates, fans and blowers, and the use of mechanical and later hydraulic drives were first introduced to alleviate the hard labor of the stokers. Today, these tasks are largely automated and sensors monitor every operating detail.

Regulation and Controls

Almost all operating parameters (action of drives, position of valves, temperatures, pressures, flows, levels) are registered continuously, for every part of the plant, as well as all relevant emission parameters (O_2, CO_2, CO, H_2O, SO_2, HCl, NOx, TOC, dust, etc.) so that all incidents can be carefully analyzed, even months post factum. Computer screens synoptically present information on storage and feeding, and on the operation of furnace, boiler, boiler feedwater treatment, steam turbo-alternator and condensers, residues extraction, air pollution control techniques, and forced and induced draft fans. Control systems are quite sophisticated and directly influence draft, furnace temperatures, and the position of the fire.

Combustion control follows complex algorithms, developed to ensure the right operating conditions, regarding temperature, pressure, airflows, etc.

Conclusions

A furnace is to achieve adequate control of air supply and draft, and thus of all major combustion conditions (temperature, time, turbulence) and emissions. Typically, combustion is conducted at more than 850°C, a residence time of combustion products in the gas phase of at least 2–3 s at these 850 C (or higher), and adequate turbulence to render these reasonably homogeneous. A minimum level of oxygen (e.g., 6 vol.% in MSW incineration) may also be specified, either by legal codes or by good practice. Ideally, combustion proceeds at a pressure slightly below atmospheric, so that combustion products do not spread to the surroundings, through the inevitable leaks that occur in between the main parts of the incinerator plant, as well as at its appendages.

Post-combustion

The average *residence time* (s) in the combustion chamber is given by the ratio of the physical volume of this combustion chamber (m^3) to the volumetric flow (m^3/s) at the furnace conditions (temperature, pressure) prevailing, and determined, e.g., at the combustion chamber exit. The real *residence time* follows a distribution determined by internal flow conditions, including short-circuiting and dead zones. Such distributions are rarely established, whether by computer fluid dynamics, or by tracer experiments, as described in [67].

Combusting gas leaving the primary combustion chamber is still at about the average temperature of this chamber, i.e., typically >850°C, yet its burnout must still be completed. There are several physical and chemical reasons for this, e.g.:

- *Residence times* in the (primary) combustion chamber are rather short, to make the best use of intense combustion and the concomitant high temperatures. Too large combustion chambers operate at too low combustion temperatures, causing incomplete combustion. Conversely, too small combustion chambers operate at too high combustion temperatures, causing severe slagging of refractory walls, unless these are adequately cooled, as well as thermal NOx. Cooling of such furnace walls is technically possible by integrating the combustion chamber into the boiler, or by blowing part of the secondary air through channels prepared in the refractory walls.

- Part of the combusting gas *short-circuits* parts of the (primary) combustion chamber, so that its *real* residence time is only a fraction of the *average* residence time. Hence, it is advantageous to promote plug flow by judicious selection of furnace dimensions, make use of any constructive features promoting plug flow and avoiding dead zones, and testing the resulting furnace designs by CFD.
- In zones of intense combustion, local or even general *deficiencies of oxygen* are likely to occur, either permanently, or only in case of fast, flaming combustion of unusually large lumps of waste. As a consequence, incinerator furnaces should be fed steadily, yet in small unit doses.
- Very high combustion temperatures lead to the partial dissociation of major combustion products, such as

$$CO_2 \Rightarrow CO + \frac{1}{2}O_2 \qquad (4.1)$$

$$H_2O \Rightarrow H_2 + \frac{1}{2}O_2 \qquad (4.2)$$

From chemical reaction theory it follows that the best results are obtained under *plug flow* conditions. Theoretically, these can be approached by a sufficiently large number of combustion chambers. In practice, such an ideal situation can be strived for by:

- Separating the combustion chamber into a main, primary chamber, followed by a secondary and possibly third chamber. This secondary chamber is in general use when incinerating, e.g., hospital waste in the sequence (1) primary partial oxidation chamber, yielding incomplete combusted fumes, and (2) secondary post-combustion chamber fitted with an auxiliary burner for raising the temperature and provisions for generating swirl and thus promoting complete combustion.
- Conventional combustion chamber (e.g., featuring a mechanical grate stoker), followed by a zone of highly turbulent mixing, produced by the injection of high-speed secondary air.

Total organic carbon is a lump parameter of flue gas organics, measured off-line by means of flame ionization detectors and expressed as mg CH_4-equivalent per Nm^3. Detailed identification is both seldom conducted and tedious, yet of possible interest in a larger environmental debate, or for dedicated monitoring of POHC (principal organic hazardous constituent) during test burns of hazardous waste [68, 69], e.g., at the Incineration Research Facility (IRF). US EPA monitored the environmental performance of hazardous waste incinerators by ordering test burns to be conducted. The legal framework is described in [70].

Under controlled laboratory conditions Dellinger et al. applied the gas-phase thermal stability method to rank the incinerability of 20 hazardous organic compounds, selected on the basis of frequency of occurrence in hazardous waste samples, apparent prevalence in stack effluents, and representativeness among hazardous organic waste materials. Their major findings were [71]:

- Gas-phase thermal stability is effective in ranking the incinerability of hazardous compounds in waste.
- Numerous PICs were formed during thermal decomposition of most of the compounds tested.
- A destruction efficiency of 99.99% is achieved after 2 s mean residence time in flowing air at 600–950°C (Table 4.1).

Conclusions

Post-combustion is essential because primary combustion chambers are too limited in residence time and in mixing and homogenization capabilities to ensure steady burnout reliably and permanently. Post-combustion is preceded by a zone of intense mixing, to homogenize oxygen-rich with oxygen-lean strands; it proceeds as long as temperature remains above, say, 500°C. As temperature decreases, all reaction rates tend to fall.

Below 500°C, oxidation may proceed further in case the remaining PICs can be adsorbed and converted catalytically.

The advent of selective catalytic reduction (SCR) paved the way for organized oxidation of PICs, the semiconductor catalysts used being capable of (first) NO reduction and (second) semi-volatile PICs (PAHs, dioxins) oxidation, even at temperatures of only 200°C.

Heat Recovery

The sensible heat contained in flue gas can largely (thermal efficiency η_{Therm} typically 75–85%) be recovered in waste heat boilers. Normally, medium-pressure (1.5–4.5 MPa) boiler operation is favored, to avoid high-temperature super-heater corrosion problems. Fly ash is often tacky above 600°C; hence the contact surfaces are preceded by radiant cooling surfaces. These are specially designed for:

- Limiting adherence and deposition of hot, tacky particles
- Convenient cleaning (rapping of boiler tube panels, soot blowing, shot cleaning of tube banks)
- Easy inspection

During a furnace standstill, it is advisable to keep the boiler tubes hot, by means of imported steam, in order to avoid corrosion by hygroscopic acidic deposits, such as chlorides. The same holds for flue gas cleaning plants.

Table 4.1 Processes influencing upon the formation of products of incomplete combustion in mechanical grate municipal solid waste incinerators, factors of influence, possible remedial action, and influence of the 850°C, 2 s Rule

Nr	Process	Factors of influence	Possible positive action	Influence of the 850°C, 2 s Rule
1	Drying	Heat radiation Early ignition of high-calorific materials	Mix dry and wet waste Preheat air Use a reverse reciprocating grate (mixing)	May be mildly positive, without exerting much direct influence
2	Heating and Ignition	Radiating Heat Ignition of adjacent materials	Noncritical process	Almost none
3	Thermal decomposition	Material Type Temperature Heat supply rate	Premixing refuse Poking and mixing action of the grate	None
4	Flaming combustion	Rate of thermal decomposition Supply of air	Adapt air distribution along the grate Enrich with oxygen	May be mildly negative, by requiring a hot furnace operation
5	Mixing the gases	Furnace geometry position and diameter of air injection nozzles	Improve the design to increase turbulence Injection of more high velocity secondary air	None
6	Post-combustion	Contact time Temperature	Apply the 850°C, 2 s Rule	Important
7	Avoidance of soot formation	Correlated with (3), (4), and (5)	As for (3), (4), and (5)	None

Plants Without Heat Recovery

In small or batch-operated plants, flue gas is cooled by injecting quench water in a cooling tower surmounting or following the furnace, or by admixing cooling air [9]. These methods increase the gas flow at standard temperature and pressure typically by 30–50% for water injection and by 300–400% for admixing air, which quite considerably inflate investment and operating costs of the gas cleaning plant.

In large-scale incinerators, *heat recovery* using either waste heat or integrated boilers is the most appropriate for cooling the flue gas prior to its cleaning, provided that the steam generated can be used for in-plant or other useful purposes, such as power generation, district heating (winter) and cooling (summer), water desalination, sludge drying, vacuum generation, etc. Still, such heat recovery proceeds under adverse conditions (corrosive and fouling flue gas), requiring considerable investment and diminishing plant availability.

Generated revenues and avoiding the extra cost of requiring much larger gas cleaning plant may offset these disadvantages. Moreover, since heat recovery is a more sustainable option, recovery may be mandatory, even regardless of economic factors.

Boiler Design

The design of a boiler mainly depends on steam quality (boiler pressure + superheat temperatures), water circulation requirements (MSWI boilers feature natural convection), and flue-gas characteristics (corrosion, erosion, and fouling potential). When selecting steam parameters for waste fired boilers, a compromise is searched between yield of power generation and superheater lifetime: an operating pressure of ca. 40 bar (4 MPa) and 400°C are common choices when power is generated [9].

Corrosion becomes more severe, as steam temperature increases. Steam superheaters are especially vulnerable: since they operate at the highest temperatures of the steam circuit they are located at the high temperature side of flue gas and boiler. Moreover, their internal cooling is of low grade (medium pressure steam, stead of boiling water). Corrosion-resistant materials and coatings are key in increased conversion efficiency and reduced maintenance in waste-to-energy (WTE) plants. Another possibility is to heat the steam superheater in a separate natural gas or oil-fired furnace, an option first tested at Moerdijk, the Netherlands.

During the 1960s, boilers were designed according to conventional rules: compact construction and a high rate of heat transfer, sustained by relatively high linear gas velocities. This design was at the source of failures: some superheaters, designed for 20,000 operating hours, barely reached 3–4,000 h. Linear gas velocities selected for high heat transfer rates also create conditions leading to rapid fouling or even complete clogging of entire tube banks and to rapid corrosion [72–74]!

From the 1970s, some simple rule-of-thumbs emerged that led to the design of large-volume, less efficient boilers, however, without the operating problems cited afore:

- Convection surfaces in the boiler passes are placed only after 1, 2, or even 3 empty boiler passes, so that the flue gas temperature is lower than 600°C or at most 650°C. In this temperature range, fly ash is no longer too tacky thus less fouling.
- The clearance between superheater tubes is wide and the approach velocity is low (only few ms^{-1}) limiting inertial fly ash deposition.
- Deposited fly ash is periodically removed using steam jets or dropping shot onto tube banks.

Chlorides, chlorine, and hydrogen chloride play an important role in some forms of corrosion. Yet, also other factors play a synergetic and decisive role, often related to the creation of electrochemical cells with on one side tube metal, on the other the tube deposits. Rate controlling is the electric conductivity of the deposition layer, not the amount of chlorine in the system. Basically, the presence of molten phases on the tubes must be avoided. Rasch studied the thermodynamics of the formation of these phases in some detail [75].

Corrosion Problems

Most gases attack plain steel. Combustion of MSW generates a highly corrosive environment composed of combustion gases and ash and laden with HCl, SO_2, chlorides, and (subsequently) sulfates. Corrosion rates rise with temperature and – depending on metal structure and composition – diminish by formation of protective layers. Coherent consideration of corrosion processes is difficult, as physical, chemical, operational, metallurgical, and crystallographic parameters interact and the precise origins of corrosion vary from case to case, are multiple, and generally difficult to identify. Thermodynamically speaking, some extent of corrosion is unavoidable. Countermeasures may help to reduce corrosion damage to acceptable levels. These require both constructive and operational counter-measures. Low steam parameters in the boiler system, long residence and reaction times (for preliminary sulfatation of chlorides) before entering in contact with convective heat surfaces, lowering the flue-gas speed, and leveling of the speed profile may all be successful. Protective shells, tooling, stamping, and deflectors can also be used to protect and safeguard heated surfaces. A compromise must be found in determining the boiler cleaning intensity between best possible heat transfer (metallic pipe surface) and optimal corrosion protection [76–79].

Currently, corrosion phenomena are observed on superheater tubes particularly. The key role of formation of a molten phase is obviously associated with ash composition and flue gas temperature. The deposit morphology is related to the flue gas flow pattern, to the mechanisms of corrosion and corrosion rates. A theoretical analysis and enumeration of corrosion's numerous forms and appearances are given in the EU Reference Document on the Best Available Techniques for Waste Incineration [38].

In the 1950s and 1960s, Germany built numerous large MSWI plants. Refuse was assimilated to fuel free of charge and the first generation of plants was designed to squeeze maximum power from this resource. Soon, severe corrosions were encountered and their sources were analyzed; several major areas of concern were identified [38, 72–79]:

- Severe corrosion occurred in integrated boilers, affecting mainly the lower half of the boiler tubes surrounding the combustion chamber. This form of corrosion derives from alternating oxidizing and reducing conditions, which prevent protective and coherent oxide films to form. It proceeds through formation of $FeCl_2$ in an oxygen-deficient flue-gas atmosphere, e.g., below oxide films, tube contaminations, or fireproof material especially in the furnace area. $FeCl_2$ is sufficiently volatile at these temperatures to be mobilized. An indicator for such conditions is the periodic appearance of CO. Corrosion products appear in flakey layers. Today, this part of the boiler is clad with protective refractory, often thermally conductive silicon carbide.
- High-temperature superheater corrosion. Corrosion occurs in synergy with other factors, such as inapt boiler design and the accumulation of tacky deposits on the superheater tube banks. Hydrogen chloride and chlorine play a major role in an electrochemical system constituted by boiler and especially superheater tube deposits: hydrogen chloride is released by conversion of alkaline chlorides into sulfates, and attacks iron. Corrosion is observed in MSW incinerators with flue-gas temperatures >700°C and at pipe wall temperatures above 400°C. The corrosion products are black, firmly bonded, and include red hygroscopic $FeCl_3$.
- Molten salt corrosion. Flue-gas contains alkali salts, which form low-melting persulfates (Na- and $K_2S_2O_7$) and various eutectics. Such molten systems are highly reactive and cause severe corrosion or even react with the refractory lining and destroy it mechanically.
- Standstill corrosion creates problems mainly after a shutdown, whether scheduled or accidental. $CaCl_2$ deposits are hygroscopic and show deliquescence, whereas some heavy metal chlorides may even hydrolyze, liberating free HCl. Electric tracing is required to keep such deposits dry during standstill periods.
- Dewpoint corrosion is associated with acid gases that condense at the cold, rear end of the boiler. Temperatures below 110°C may suffer from HCl condensation; sulfuric acid may even condense below 160°C.
- Superheaters may suffer damage from erosion due to excessive flue gas approach velocities and/or excessively strong soot blowing. Such soot blowers are difficult to adjust: if the jets blow too hard they cause erosion, if too soft, soot blowing is useless. Specialized services now blast deposits by appropriate use of explosive charges.

Sulfatation

Salts and metal chlorides sublimate at furnace temperature, leaving bottom ash as a cleaner residue [75]. In the first boiler passes the temperature remains still above 650°C and fly ash is still tacky. Below 600°C, flue gas may come into contact with tube banks, without excessive risk of fouling these rapidly. Nevertheless, tube deposits still form by separation of nonsticky particles, by inertia and interception. These deposits also collect chloride salts that de-sublimate and condense. Thermo-dynamically, most chloride salts are no longer stable, as they were at furnace temperature. Upon contact with SO_2 they gradually convert into sulfates by generic reactions such as:

$$MeCl_2 + SO_2 + \frac{1}{2}O_2 = MeSO_4 + 2HCl$$

$$(Me = bivalent\,metal)$$

(4.3)

Such reactions also consolidate and harden deposits. Moreover, while liberating HCl they contribute to corrosion processes: HCl slowly oxidizes to Cl_2 that diffuses to the tube metal and attacks it; after it is reduced to HCl the same corrosive cycle starts over. From this viewpoint, it is favorable that the flue gas is rich in SO_2 and that sulfatation proceeds before the salt-laden fly ash deposits on the tubes.

Flue Gas Composition

Flue gas composition is determined by several factors of influence. The most important one is waste composition: all entering elements will also leave the plant, whether as flue gas or as solid residue. Mass balances, together with waste composition data, allow estimating the flue gas and the residue composition, even though some assumptions are needed regarding the distribution of the relevant elements over the various output streams. A second factor is the technology used: mass burning of MSW yields much more bottom-ash (typically 20–30 wt.% of MSW) than fly ash (2–3 wt.% of MSW). Fluid bed incineration of the same MSW will turn this relation in favor of fly ash, which may reach, e.g., 10–12 wt.% of MSW. As a consequence, the coarse fraction of fly ash will be less contaminated, following an effect of dilution by bed material and other fines. A third factor is related to the operating conditions used: lower flow and velocity of primary air reduces the entrainment of fly ash and also leads to higher bed temperatures and hence to more sintering of ash and to more volatilization of various heavy metals, e.g., Cd, Cu, Pb, and Zn, that eventually de-sublimate onto the fly ash.

Flue gas composition is also influenced by the excess air amounts practiced: primary air activates the fire in the combustion zone, yet cools the furnace in the drying and burnout zones; excess secondary air merely dilutes the flue gas.

To avoid willful dilution with ambient air (to make concentration figures look lower), analytical data are generally expressed at some standard concentration of either oxygen (e.g., 11 vol.% O_2) or carbon dioxide (e.g., 6 or 8 vol.% O_2). Similarly, the concentration of obnoxious compounds is generally expressed on a dry gas basis.

In modern plants, numerous parameters are monitored continuously, e.g.:

Oxygen, on the basis of its paramagnetic properties, or using semiconductors reacting to the oxygen concentration.

Carbon dioxide, water vapor, sulfur dioxide by Fourier-transformed infrared (FTIR) absorption

Hydrogen chloride and fluoride

Nitrogen oxides

The residue composition also depends on the partition between bottom ash, boiler slag (only a small amount), fly ash, neutralization residues, and fine dust and aerosols that escape uncollected. Numerous studies have considered such issues.

Dioxins

More than a century ago dioxins first drew the attention, while their synthesis afflicted laboratory workers with chloracne. The same happened after isolated incidents in chemical industry, e.g., Monsanto at Nitro, BASF at Ludwigshafen, or Philips-Duphar at Amsterdam. A much more spectacular accident occurred at Seveso (N. of Milan): after a run-away in a herbicide synthesis reactor, its contents were vented all over Seveso, causing trees to lose their leaves, death to various animals, as well as the evacuation of 10,000 inhabitants (1976). People exposed to dioxins are still being monitored today, to detect any eventual symptoms or mortality. Epidemiological investigations show the appearance of rare, soft tissue cancers and neurological afflictions, yet no net increase in mortality (cfr. Public Image).

Dioxins were discovered on MSW incinerator fly ash in 1977 [80]; it took some 15 years more to recognize as major sources several processes in iron and steel industry, as well as in the melting of metal scrap. Dioxins have been at the center of enormous efforts, first to develop, standardize, apply, and ameliorate analytical methods and determine potential dioxin sources as well as possible pathways to formation, then to try and meet the extremely low emission limit values during everyday operation [81–86].

Details of the mechanisms forming dioxins still today remain controversial [87–89]. Theories started with the trace chemistries of flame (Dow Chemicals Co.), continuing with various precursor theories (many researchers) and culminating with the de novo theories, worked out in most detail at Forschungs-Zentrum Karlsruhe. In the first theory, dioxins are inseparable from any combustion process [90]. Precursor theories focus on chemical, often catalytic conversion of dioxin-related structures [91–93], such as phenoxy radicals, chlorophenols, chlorobenzenes, polychlorinated biphenyls (PCBs), and also polycyclic aromatic hydrocarbons and related structures, converting into dioxins. Finally, de novo theory is based on a low-temperature catalytic conversion of almost any carbonaceous structure, such as soot or its various precursors, into dioxins and furans, or PCDD/F [94–98].

Several pathways lead to dioxins [109], yet their relative importance, as well as the precise nature of the catalysis at work will always remain elusive in each particular reactive system. Moreover, there is no mutual exclusion between pathways. Much attention was also given to metal catalysis in dioxins formation [99–102]. Other work related to the prevention of dioxins formation [103–105] or its destruction in fly ash [106–108]. Early and current abatement of dioxins from flue gas is covered in [109–111].

Dioxins in Incineration

During several decades, incinerators have formed the major source of dioxins emissions.

Strangely enough, they were also destroying dioxins, namely, those entering the furnace together with the MSW [112]. Dioxin balances have been established several times in the 1980s, showing that the input and output of dioxins in the plant was similar, yet not necessarily the dioxins fingerprint, i.e., the distribution of various isomer groups and congeners.

Although dioxins are considered to be extremely environmentally stable, they do not survive the combustion process. So, more than 99% of the dioxins entering are destroyed. At the entrance of the furnace and even after the practical end of active post-combustion, no dioxins can be found; at most their basic structures are present [112–115].

Rapid dioxin formation occurs once the flue gas attains a window between 400 and 250°C. A rate maximum of formation occurs at 300–350°C [96].

Explanations differ, yet it seems accepted that the formation is a catalytic process, so that particles play a role, whether suspended in flue gas or deposited from it. Oxygen is required, probably to reactivate the catalyst, after it is reduced while chlorinating aromatic and aliphatic structures.

Salient Factors in Dioxins Formation

Dioxins formation is affected by quite a large number of significant factors, subdivided into two groups: first, operational factors, second, related to chemical, composition, and catalytic factors, such as catalysis, carbon, oxygen, water vapor, and chlorine [116]. Each of these has several impacts, often with various mutual interactions and it is unlikely that their ranking and relative importance under varied conditions in diverse systems will ever be established once and for all. An intrinsic difficulty in studying dioxins formation is a matter of timescale: the occurrence of a combustion setup, start-up, or shutdown has a certain timescale [117], yet that of dioxins may follow hours, days, or even weeks later (memory effects) [85, 118]. Several factors explain such memory effects: dioxins form from fly ash deposits slowly, and even slower in lower deposit temperatures. In some cases, there may be chromatographic effects, semi-volatiles such as dioxins getting adsorbed and desorbing again later. Wet scrubbers made of plastic dissolve dioxins during upsets and start-ups that desorb again into clean gas later [119].

Incinerator operating factors are of paramount importance. Poor combustion conditions may result from "bad" waste, i.e., either too poor (low temperature) or too rich (excess evolution of volatile matter). These "bad" operating conditions not only lead to more PICs and PAHs (a small fraction of which converts into dioxins), but also to a prolonged increase in dioxins (memory effects: PICs adsorb on boiler deposits and continue generating dioxins afterward). Poor combustion conditions result often from feeding too much at a time, without adequate premixing wastes of different origins and quality. Combustion upsets are notable by a development of peaks of carbon monoxide accompanied by total organic carbon (TOC), a measure for the amount of PICs present. Combustion conditions may be improved by both technology (grates, furnace geometry) and operating skills (mixing and feeding waste, providing primary and secondary air). Nevertheless, firing fuels such as MSW always bring in a factor of chance. With respect to dioxins, the following factors may help:

- Firing well-mixed, homogenized waste only. Humidity transfer from moist vegetal waste to paper and board and dispersion and mixing of high-calorific waste (plastics and rubber) in the bulk of MSW are positive factors, i.e., prolonged storage and periodic mixing of the bunker's content, or mixing moist garden waste with high-calorific commercial waste.
- Using low rates of primary air. This reduces the amount of excess oxygen in the flue gas, as well as the entrainment of dust particles, which is a source of dust deposits and of boiler fouling and corrosion.
- Steady combustion conditions. No large packs of high-calorific waste taking fire together.
- High-quality mixing of gases at the furnace exit.
- Ample post-combustion chamber volumes, at adequately high temperatures and mixing levels.
- Designing post-combustion volumes by means of computer fluid dynamics, for good mixing and avoiding short-circuiting as well as dead zones.
- Limiting residence times in a temperature window ranging from 500°C down to 200°C.
- Operating electrostatic precipitators, at low temperature, not more than 200°C, by extending waste heat boiler surfaces and limiting boiler fouling.

- Avoiding building up and extending deposits on boiler tubes, collection plates in electrostatic precipitators, in flues, etc., by limiting the approach velocity.

The quality of operation can be judged by the permanent absence of CO- and TOC-peaks.

Ideally, their frequency should be nil on a daily basis. Should such peaks still occur, they can be termed "very serious" (CO = 10^3–10^4 mg/Nm^3), "serious" (10^2–10^3 mg/Nm^3), or "benign" (10–10^2 mg/Nm^3). TOC-peaks concur with CO-peaks, yet their height and width differ. The reason for such short-lived peaks is either overfeeding (too much at a time) of fluid beds, or inadequate mixing of MSW fed to mechanical grate units [42].

Complete combustion, mixing of flue gas by blowing in secondary air at high speed, and absence of setups are all primordial operating factors; definitely less dioxin is formed in case excess oxygen is limited.

Another important operational domain is related to the cooling of flue gas: fast and deep cooling limits dioxins formation. Slow cooling of flue gas, in contact with deposited dust, has an opposite effect. For small plants, e.g., metal foundries, quenching off-gas is a suitable prevention measure.

Dust removal takes dioxins away, since these semi-volatiles report to fly ash, especially at low temperatures. Baghouse filters are designed to clean gas down to the very low dust levels required to reach the level of 0.1 ng TE/Nm^3. Any imperfections should be observed by means of tribo-electric sensors, opacity measurement, providing immediate warning in case of dust breaking through.

Chemical and Catalytic Factors

A cardinal chemical factor is related to the presence of transition metals providing the catalytic effects required to fix chlorine on carbon structures and also to oxidize the latter so that dioxins are liberated, together with scores of other surrogate and precursor compounds [47]. Catalytic metals are likely to be associated with particulate, in particular its finest fraction. The latter absorbs the de-sublimating metal salts (Zn, Pb, Cu, Cd, etc.) condensing after having been volatilized at flame temperature [67]. Copper is obviously a premium catalyst; it is often better represented in fly ash from fluidized bed units than in that from mechanical grate units [42]. This could be due to erosion effects, affecting copper wire. In China, fly ash is much leaner in heavy metals than in the EU. Another catalytic substance is iron oxide.

Mixing fly ash with inert materials and carbon creates de novo, dioxin-generating activity. Matrix effects and its particulate carrier are important [120], so is the supply of oxygen to the system: after chlorinating carbon or oxidizing carbon structures, the catalyst is in its reduced form. Oxygen restores a higher valence, required for reactivity. The relations between carbon structure and dioxin formation are still all but elucidated. The presence of the element chlorine is essential in dioxin formation, yet chlorine is ubiquitous in incineration. Factors of influence are numerous and their effects are manifestly complex, interdependent and difficult to pinpoint! Dioxins formation has been studied at full plant level [112], at pilot scale [113–115], and at laboratory level [121, 122]; it was simulated by CFD [123]. Thus, the discovery of dioxins eventually has prompted enormous research efforts, with the fortunate result that incineration became a much more controlled technical process and that the cleaning of flue gas became much deeper (cfr. Tables 4.2 and 4.3).

Flue Gas Cleaning

In MSWI flue gas a deep cleaning is essential. Public and political pressures have been so powerful that MSW incineration is at present the most regulated and best controlled form of combustion. Flue gas cleaning addresses successively [30, 37]:

– Particulates and dust, including the associated heavy metals
– Acid gases, such as HCl, HF, and SO_2
– Nitrogen oxides such as NO, NO_2, and N_2O
– Semi-volatile organic compounds, such as PAHs, PCDD/Fs (dioxins), and PCBs

Yet, the precise composition of the flue gas cleaning train depends on numerous options that can be combined in a large variety of flue gas cleaning schemes. Most existing plant during the 1980s and 1990s were forced to revamp this train at least once or even several times, leading to redundancy in the ways these various duties are addressed, e.g.,

– Baghouse filters were often added at the tail of the plant, to complete the preliminary separation by a preexisting electrostatic precipitator; in other plants the ESP was scrapped, because of redundancy and the formation of dioxins at high ESP operating temperatures.
– Dry acid gas scrubbing was supplemented at times by semi-wet or wet units.
– Activated carbon adsorption retains semi-volatile organics that eventually would be destroyed during selective catalytic reduction of NOx.

A survey of best practicable options is given in [37].

HCl is an acid, irritating gas, yet it is eminently soluble in water and thus easily scrubbed out from flue gas (together with HF and HBr, both present at about 100 times lower concentration levels). The resulting diluted solution can be distilled to yield a commercial concentration. Yet, HCl is not in high demand and sales may require removal of trace organics as well as iron. An alternative is using it as a leaching agent, to remove heavy metals from fly ash. In case such recovery options are not followed, yet, the acid needs to be neutralized, e.g., by means of lime.

Chapter 5
Selection of Incinerator Furnaces

Selection Criteria

The *selection* of a particular type of *furnace* mainly depends not only on the type(s) of waste to be incinerated (which also determines the possible feeding methods), but also on numerous other factors, such as plant capacity, the operating schedule required, heat recovery, the amount of ash to be handled, and also its physicochemical nature and softening point, etc.

Off-gases and *liquids* are relatively easy to handle using an adapted burner in a simple, tailored combustion chamber, but the incineration of solids, sludge, and paste... may take place under a wide range of combustion conditions and in different types of furnaces.

Furnace types can be classified, according to:

- The contact of waste with combustion air (i.e., in co-current, counter-current, or cross-current relative flow; mechanical and pneumatic agitation, etc.)
- The degree of filling the combustion chamber with solid material
- The choice made between *dry* ash and slag melting conditions (so-called wet-bottom furnaces (not to be mistaken for dry or wet (in water) extraction of combustion residues).

Possible *plant capacity* may be limited by either construction methods, or experience factors; e.g., for mechanical grate at typical capacity 2–20 Mg of MSW/h or rotary kiln furnaces (typically 0.5–5 Mg of waste/h there is only limited experience available once a given size is exceeded. Higher capacity is achieved by providing parallel lines of generally identical capacity and make. Spreading capacity over two or more lines also allows more flexibility, in case of shutdown of one train or of variable supply of waste. Extrapolating existing units to an untested scale may lead to unexpected problems in thermal units. Such was the case in the 1970s

This chapter was originally published as part of the Encyclopedia of Sustainability Science and Technology edited by Robert A. Meyers. doi:10.1007/978-1-4419-0851-3

for Monsanto's Landgard partial oxidation plant at Baltimore, the Andco-Torrax gasification plant at Leudelange, and the Occidental Petroleum Garrett Pyrolysis plant at El Cajon, Ca. [9, 21, 22].

Heat recovery often features a separate *waste heat boiler*, consecutive to the combustion chamber. Waste with high HHV may also be fired in a furnace, integrated into the boiler structure (*integrated boiler*). The ceiling and sidewalls of the combustion chamber are structurally formed from vertical and inclined boiler tube panels constituted from parallel finned tubes welded together. The tubes are covered by studs sustaining refractory mass, rammed onto the tubes so as to protect them from fouling and corrosion [9].

Pollutant control at times may decide upon the type of furnace to be used or on its operating conditions. Sulfur dioxide (SO_2) is easily captured in a fluidized bed combustor, operating at 850°C, which is the optimal temperature for reacting SO_2 with lime or limestone. Similarly, thermal NOx can largely be avoided at that temperature. Nevertheless, there is always a negative correlation between two types of pollutants: NOx on the one hand and CO + TOC (or PICs) on the other. In case fuel-NOx problems are expected the technique of *staged combustion* is used, which is composed of two steps:

1. Combustion conducted with a deficiency of air (first step, at high temperature), thermally reducing fuel-NOx
2. Post-combustion with ample air and at low temperature

In most cases, this technique will alleviate the problem. Combustion conditions also fix two important factors: (1) ash tends to sinter, soften, and eventually melt, as temperature rises, and (2) the distribution between fly ash and bottom ash also evolves with temperature. Other cardinal factors are the presence of oxidizing or reducing conditions and of halogens, sulfur, etc. [124].

Small-scale incinerators (capacity <2 Mg waste/h) were often operated in a one- or two-shift schedule, but today continuous operation is always to be preferred, since it enhances useful capacity and reduces auxiliary fuel requirements during start-up, thermal wear on refractory, and plant emissions.

Start-up and *shutdown periods* are much more polluting [85], and there is a strong tendency to allow only pure auxiliary fuel to be burnt during these periods. Waste firing can only start once the operating temperature is reached.

Simple, Small-Scale Forms of Incineration

Burning in the open, e.g., in a dedicated open pit, a barrel, or the foot of an old stack, is both highly polluting and difficult to master technically. An open pit burner was developed by DuPont to incinerate waste [35]. Wigwam or tepee conical burners were used for burning trash, mostly in remote communities [35], in a more controlled manner than is feasible in the open. American apartment buildings used at times to be equipped with chute fed incinerators [125, 126].

These practices should be banned from any densely populated area. Still, in remote areas, e.g., in parts of the USA and Canada, it is often considered the only option practicable and these units are still largely advertized in the USA, even though their use would probably be forbidden in the EU and Japan. Open burning is obviously a major source of pollutants [127–129].

Stationary Furnaces

Summary. Simple, stationary furnaces are in general use for firing gaseous and liquid fuels or even solid waste, on fixed or rotary grates.

Principles. A furnace combines several essential functions, namely:

- Limiting the cooling of the flames and sustaining an adequate furnace temperature
- Providing adequate retention time in the combustion chamber
- Preventing the uncontrolled entrance of air
- Organizing the flows of incoming primary and secondary air and outgoing flue gas, without undesirable dead corners, entries of false (uncontrolled) air, or diffuse spreading of fumes in case of a temporary rise in furnace pressure

Avoiding *smoke* spreading around requires operating at slightly subatmospheric pressure, since incinerators are always somewhat leaky, a consequence of the heating and cooling cycles inflicted upon refractory and casing. For this reason, furnaces formed from welded membrane steel or boiler tube panels are very popular, ever since their first introduction ca. 1970. The selection of waste burners, their position and capacity, flame orifice, air supplies, mixing, and thermal buoyancy characteristics are prime factors determining performance. The mixing characteristics of the furnace are enhanced by appropriate injection of secondary air, enhanced back-mixing of flue gas, created by reducing the cross section of the outlet and by providing periodic changes in the direction of flue gas flow.

Heat release rates are high when burning high-calorific gases or atomized hydrocarbon liquids; they are much lower when burning sludge or wastewater. Where required, a separate post-combustion chamber is used to control PICs, soot, or smells, with its temperature controlled by an auxiliary burner. An alternative is to provide a catalytic post-combustor [130].

Construction and operation. Stationary furnaces refer to a plain combustion chamber, either horizontal or vertical, of a cylindrical shape or box-type, and fitted with the required start-up and auxiliary burners. Horizontal tubular furnaces are most common, possibly aligned with equally tubular waste heat firetube boilers. Box furnaces exhibit dead corners and were used less than half a century ago.

Vertical furnaces occupy less floor space, gradually narrowing to form the stack and deriving draft from this geometrical design. Nowadays, this arrangement becomes less common, since incinerator flue gas generally requires stepwise and multistage cleaning.

Applications. The stationary furnace is used for burning gaseous and liquid waste flows, including off-gases, solvents, oils, wastewater, pumpable sludge, and meltable and paste-like waste streams. Plastics proper are difficult to fire through a burner, for liquid burners will spin threads of molten plastics. Special burner designs fire several streams simultaneously, e.g., auxiliary fuel, waste oil, wastewater, and pumpable sludge. Alternatively, various wastes may be injected either into a stable flame or tangentially to it. Wastewater may be largely evaporated in a forced circulation evaporator and then radially blown into the flame of an auxiliary oil burner.

Advantages and disadvantages. The main technical limitation of an empty combustion chamber is the lack of provisions for eliminating ash or other residues. Ideally, the ash is fine and high melting and blown out of the furnace, and then separated by the air pollution control devices. Residual ash can be eliminated according to different schedules, such as:

- Operating on a daily shutdown schedule for manual or mechanical cleaning
- Periodic or continuous elimination of ash using suitable mechanical means, such as drag conveyers, augers, retractable grates, rotary grates, etc.

Larger units may incorporate rugged, resilient, yet flexible mechanical provisions to convey ash outward. Since it is undesirable that air leaks in through the ash removal system, a wet or dry sealing system is necessary.

Mechanical Grate Incinerators

Summary. Mechanical grate stokers were originally developed for coal [8, 59], yet since the 1930s they have increasingly been used for MSW.

Principles. Traveling grates support the fuel, while conveying it through the furnace, from the front feeding to the ash-discharging side. Staircase grates provide some tumbling action, when fuel drops from one section to the next. *Reciprocating grates* feature individual grate bars, mounted on alternating moving and fixed frames or sledges; moving the sledge conveys the overlaying fuel and – upon its retreat – turns it over that resting on bars from fixed frames. Several arrangements are possible, e.g., with alternating fixed and mobile steps, or with alternating fixed and mobile staircases juxtaposed. A survey of patent literature reveals a richness of ideas to move waste and separate ash [8].

Construction and operation. Most mechanical grates are subdivided conceptually or physically into successive and distinct drying, combustion, and burnout sections, sometimes separated by small walls, where waste tumbles from one level to the next. The position of the fire is somewhat controlled by the mechanical action of the grate, which supports, conveys, and stirs the refuse during drying, combustion, and burnout. The most common types of grate are reciprocating, reverse reciprocating, roller, rocking, and traveling grates. Proprietary, patented devices provide controlled motion, poking, mixing, and sifting ash between individual grate bars.

Primary combustion air is supplied under the various grate sections to cool the grate and accelerate the burnout of the residue. Today less primary air is used, reducing dust entrainment and the flow of flue gas per unit, and improving thermal efficiency.

Air requirements for drying refuse or for burning out clinker residue are quite low, but supply is ill-adapted to real requirements when active combustion takes place. The vapor and gases, resulting from drying and heating the refuse, are rich in oxygen; combustion products evolve as hot, oxygen-deficient strands. Both should be thoroughly mixed by means of powerful jets of secondary air, blown in through high-velocity nozzles, located at the exit of the combustion chamber. After completing further combustion the flue gas is cooled by a waste heat boiler or – in small plant – by injection of water into a cooling tower. Finally the flue gas is cleaned.

Typical combustion conditions are 850–1,050°C, excess air of 80–200%, but there is a strong tendency to limit it to 6–9 vol.% of oxygen in the flue gas. Some operating conditions are specified by codes, e.g., the EU Directive 2000/76/EC:

- Minimum operating temperatures of 850°C and minimum residence time of 2 s at this temperature
- Minimum level of 6 vol.% of oxygen

Applications. The basic application of mechanical grate stokers used to be for firing calibrated coal. Calibration ensures that all lumps or particles burn out after the same residence time, i.e., by the time the coal arrives at the end of the grate. MSW, however, is all but homogeneous. Deviations from uniformity are catered for by providing a feed that has been homogenized and aged (moisture transfer) in the MSW pit and by specific grate action. A number of options are available for co-firing sewage sludge, waste oil, plastic-rich fractions, etc.

Advantages and disadvantages. Mechanical grate operation has been evaluated a number of times, and in the Western society it can boast decisive advantages with respect to the numerous alternatives tested over more than a century.

Its major limitations are:

- Limited to waste that is supported by a grate. Powders, sludge, and liquid or melting waste are excluded, except for marginal amounts. Reporting to grate siftings impairs their quality.
- Less suitable for waste with extremely low or very high HHV, unless both are well mixed. Fluid bed units are much more flexible in this respect.

In southern climates and developing countries, MSW largely consists of putrescible organics and may be too moist to sustain combustion without auxiliary fuel.

Shaft Furnaces

Summary. Shaft furnaces were rather extensively used a century ago, yet they are currently unusual in waste management [9]. Their fields of potential application are briefly discussed.
   Principles. The charge is always fed on top, and slowly descends to the hearth by gravity. The air rises generally from the bottom of the unit, activating the fire in the hearth (countercurrent operation). Unless the feed is carefully calibrated, the gas preferentially rises along bigger channels, reducing the quality of contact with air, as well as volumetric capacity. For this reason, shaft furnaces were unsuccessful in tackling raw municipal solid waste; preliminary shredding markedly improved their performance.
   Vertical shaft furnaces have been operated in co-current, crosscurrent, or countercurrent (Fig. 5.1). Usually the last option is selected, since it easily materializes and ensures heat economy, the incoming charge being dried and preheated by the outgoing gas. As a consequence, any moisture and volatile matter evolving from the charge reports to the gas stream, charging it with organics, tars, and odors. Co-current operation hence has been applied in some gasifiers, with the purpose of cracking tars. Crosscurrent operation was applied by WSL/Foster Wheeler in an unsuccessful rubber tire pyrolysis process.
   Construction and operation. The shaft furnace consists of a vertical, cylindrical shell protected by inner refractory and thermally insulating lining. Top feeding features a suitable lock for exclusion of air and possibly a distributor for equal distribution of the feed over the entire cross section. Ash extraction proceeds mostly either by means of a rotary grate for ash extraction, or by periodic molten slag tapping.
   Applications. Already in Roman times shaft furnaces were used, for calcining limestone. Traditionally, they have been used in the iron and steel industry (blast furnaces), foundries (cupola furnaces for melting metals), and for wood and coal gasifiers.
   Shaft furnaces appeared in some ancient incinerator systems (1880–1930), either as combustor or as ash burnout element, the inherent heat exchange assisting in burning low calorific waste with combustion air preheated by hot ash. The rising gas is heavily charged with thermal decomposition products from waste and hence it requires post-combustion. Early Dörr, Didier, Stockholm furnaces are discussed in some detail by Reimann [4].
   Advantages and disadvantages. Major advantages are countercurrent heat exchange and a relatively low load of dust. Major areas of potential operating problems with shaft furnaces are [9, 22]:

- Volatiles and moisture emanating from the charge report to the gas, requiring either post-combustion or adequate treatment of these compounds.
- Unless special care is taken to homogenize size and shape of the feed, the rising gas will be channeling through preferential pathways in the charge and along the wall, due to both irregular bed porosity and chimney effects.

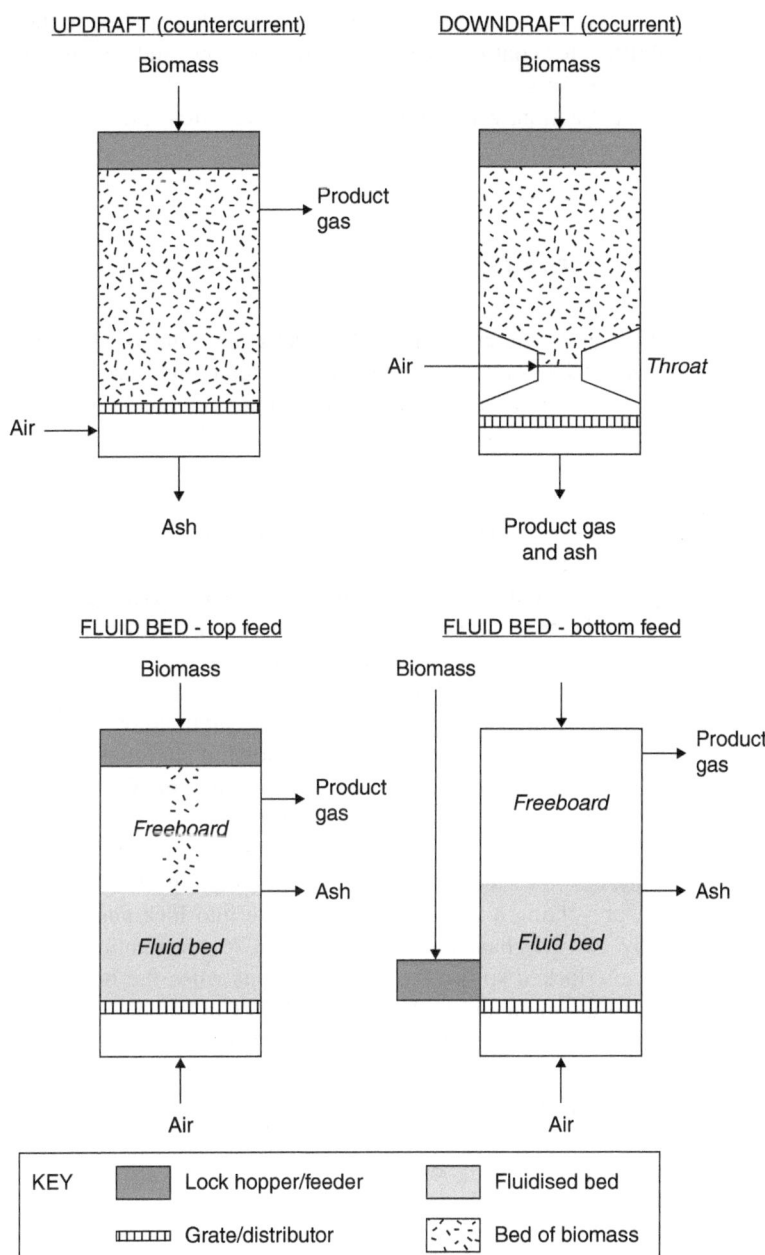

Fig. 5.1 Co-current, crosscurrent, and countercurrent operation of vertical shaft furnaces

- Peripheral fires occur, due to the larger voidage of the charge close to the walls.
- Failing possibilities for controlling temperature, gas flow, and oxygen distribution throughout the charge.
- Controlling ash extraction, whether as cinders or as molten slag.

Rotary Kiln Incinerators

Summary. Rotary kiln incinerators since the 1960s are state of the art in the incineration of industrial waste, including commercial and hazardous waste. These kilns operate either in countercurrent (cfr. long cement kilns), or in co-current (in short kilns, usual in industrial waste incineration). Cross-flow is possible only when using special constructions, e.g., involving telescopic kilns, or in mid-kiln feeding over a complex fixed feeding/rotary kiln device.

Principles. More or less as the shaft furnace, it can operate in countercurrent, co-current, or – with some more difficulty – in crosscurrent mode. Incinerator operation is generally in co-current, with either solid or liquid ash discharge. Tumbling the waste renews the furnace. Since there are no provisions for mixing gas phase strands, a post-combustion chamber is always required.

Construction and operation. A rotary kiln incinerator is typically composed of a stationary feeding system, a rotary kiln with slightly inclined cylindrical shell, a stationary ash discharge system, and a post-combustion chamber, followed by a waste heat boiler (or quench cooler), and air pollution control units [42].

The stationary feeding system consists of a feed hopper, a lock, and a steeply inclined chute. The whole fixed front panel can be mounted on rails. In a patented system, a knife rides on the sides of the feed hopper, cutting off ribbons, plastic film, and textiles, preventing a flashback of the flame into lock and hopper. The lock is formed by two mechanically, hydraulically, or pneumatically operated slides, which are interlocked so that a slide only opens when the other is closed. The lower slide and the chute are both water cooled. The feeding system can be provided with an explosion relief system, diverting a shock wave into an innocuous direction.

Several systems were tested for homogenizing feed materials and supplying them at a constant rate. Some rotary kilns were fed by a screw conveyor or hydraulically operated ram feeders. Early BASF plants used simple, strong centrifugal pumps with large clearance between rotor and housing, capable of macerating material and pumping the resulting slurry. The rotary kiln was fed from a rotating mixing and storage drum, blanketed with nitrogen.

Kiln lining. The cylindrical shell is internally lined with refractory, selected with regard to the expected operating temperature and slag melting point and reactivity. When using high-quality, dense brick, a continuous operation is necessary to avoid thermal stresses. Gradual heating up may take as much as 60 h and cooling down 24 h. A lifetime of 2 years is considered to be good in normal operation.

No general rules can be formulated regarding the best or more economic furnace lining. In some plants, inexpensive ramming compound or hard chamotte bricks were successfully used. In others, chemical attack was so extensive that lifetimes remained too short, even with expensive high-alumina or magnesia bricks. Chemical attack depends on chemical composition of both lining and ash, and on temperature. Refractory is also subject to abrasion and spalling [35].

Protecting the walls with a layer of solidified molten slag and outward cooling by water sprays have been successfully applied for lengthening lifetimes. Slag accretions can be melted away periodically by slightly elevating the temperature. Iron oxide, when burning barrels, forms low melting silicates enhancing slag fluidity. Slag reactivity and melting point has sometimes been decreased successfully by addition of suitable charges, e.g., sand.

Kiln movement. The peripheral speed of the kiln can be varied continuously using a single drive, with driving pinion and bull gear. The shell is provided with riding rings, rolling on support rollers to obtain a uniform distribution of bearing forces over the shell. In case of power failure, an auxiliary motor should drive the shell to prevent thermal deformation.

Air sealing between rotary shell and stationary loading and discharging equipment at the ends is critical. Excessive air leakage should be prevented by provision of suitable angle or segment seal rings.

In the most usual *co-current operation* both wastes and combustion air are introduced at the front end of the furnace. An auxiliary burner is installed in the fixed front panel of the kiln to enhance drying and accelerate preheating and ignition of the wastes. In the absence of such a burner, drying and preheating completely depends on radiant heat transfer from the rear part; the rate of radiant heat transfer is proportional to the fourth power of temperature (in °K), attained in the hottest part of the kiln. Operation is at 1,200–1,500°C in a slagging operating mode or below 1,000°C in the dry extraction mode.

The rotary kiln is sometimes operated in *countercurrent* when relatively wet wastes with a low heating value are to be incinerated (e.g., sewage sludge). Counter-current operation is unsuitable for other waste, because of the risk of flame flashback into the charging lock.

A *partial countercurrent* operation is sometimes used in very short kilns. A good mixing pattern is obtained by using an auxiliary burner in the fixed rear panel. The burner creates a backward gas flow along the kiln axis Fig. 5.2.

Kiln Internals

The residence time and flow pattern in principle can be modified by installing conveying spirals to guide the materiel flow, ring-dams to retain melted or chains for granular material, or by providing an enlarged cross section near the discharge end to reduce the gas velocity and provide a soaking period at high temperature.

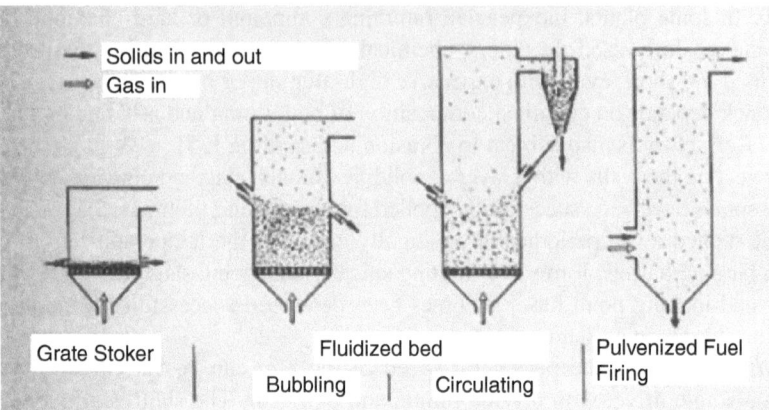

EOLSS-WASTE INCINERATION TECHNOLOGY

Fig. 5.2 Evolution from a fixed bed, over a bubbling bed and a circulating fluid bed to an entrained flow combustor, with rising gas flow (Görner [34])

A spiraling dentition can be provided in the refractory to retard the forward movement of wastes and enhance the contact between burning wastes and combustion air. The higher cost of the lining limits the practical use of these various patented devices, also prone to erosion and clogging.

Air Supply

Primary air is blown in through a set of nozzles, located on the fixed, front side of the furnace. No secondary air can be distributed along the kiln, unless it is composed of several sections of a different diameter in a telescopic arrangement. The excess of air is large, to make up for sudden variations in calorific value, and often amounts to 200–300%. Typically 8.000–12.000 m^3 of flue gas is generated per ton of waste.

Combustion air is blown in tangentially at the front end and creates a whirling movement along the wall. Superposition of the two different flow patterns results in a reasonable amount of gas phase mixing, a more uniform temperature and increased kiln capacity.

Residence time and turbulence in the gas phase are both limited. Hence, combustion is to be completed in post combustion chambers providing a supplemental residence time of 2–3 s. The temperature in these chambers is often maintained above 800°C with auxiliary burners firing fuel or liquid waste (solvents, oil).

Facts and Figures

The volumetric rate of heat generation varies, depending on the combustion temperature, between ca. 400,000 and 1,000,000 MJ h^{-1} m^{-3}. Thermal efficiency of the rotary kiln plant is low, limited typically to 55–60%, due to the large excess of air and the various heat losses.

Rotary kilns are built industrially with diameters from 1 to 4.5 m and a length typically from 3 to 15 m. The largest kilns have a capacity of 60 GJ h^{-1}. Scaling-up problems arise, because kiln volume is proportional to the square of the inner diameter, the available exposed surface of waste only to the inner diameter D_i. Hence, multiple kilns are preferred over a single, large diameter one.

Applications

The concept of rotary kiln incinerators was developed at BASF-Ludwigshafen, probably inspired by the much longer units used for producing cement clinker, for calcining limestone or for roasting pyrite and sulfide ores. Yet, the short rotary kilns used in incineration retain the tumbling action rather than countercurrent operation and intrinsic heat exchange.

Dedicated rotary kiln incinerators are capable of eliminating almost any type of industrial wastes, e.g., plastics, oil contaminated sludge, waste paint, solvents, pesticides, spent chemicals, and even explosives (in small amounts). Explosive combustion is relatively harmless, the combustion chamber being spacious and followed by a post-combustion chamber. The rotary kiln is not highly regarded as an incinerator of municipal refuse because of excessive wear of the lining and the absence of possibilities for longitudinal air distribution.

Solid and paste-like wastes, sometimes even complete barrels filled with waste are introduced into a hopper, with a lock system and a chute, located at the stationary upper end of a slowly rotating, slightly inclined cylindrical furnace. The wastes slowly slide and tumble by the rotary movement of the kiln; this provides for mixing and a periodic surface renewal of the burning charge. On their way from the higher feeding side to the lower discharge end, the wastes are rapidly dried, heated, and ignited under the action of radiant heat from the furnace walls. The kiln is generally filled up to 10–20% of its volume. The *residence time* of solid and paste-like waste depends on the length of the kiln, its speed of rotation, the possible presence of a profile in the refractory lining, and gas velocity. Generally residence times of less than 1 h are selected. The ash is discharged into a water bath, located under the lower end of the kiln. In some plants larger pieces of residue are retained on grizzly screen bars, to protect the ash-discharging conveyor [131].

The rotary kiln is used as a drying furnace for, e.g., sewage sludge, in the roasting of pyrites and sulfide ores, the calcination of limestone and the production of cement clinker. In Great Britain, Belgium, and Germany, pulverized refuse was used as a supplemental fuel in coal-fired kilns. Later, rubber tires and hazardous waste in numerous plants became a standard supplement in cement clinker manufacturing.

Westinghouse proposed a unique combination of a rotary combustor with an integrated boiler (O'conner) [35]. In only few cases wastes have been treated or incinerated in a metal-walled rotary kiln having no refractory lining at all (the red factory, of Prayon, Engis, Belgium).

Advantages and Disadvantages

The following problem areas have been identified:

- The charging chute is exposed to heavy wear because of feed sliding and tumbling and condensation of corrosive vapors. Sometimes cracks occur along the welding.
- The kiln lining is exposed to heavy wear and chemical attack. The action of corrosive melted slag is important at the kiln end mainly.
- Air sealings are exposed to dirt, wear, and high temperatures.
- The lower part of the combustion chamber is exposed to attack by entrained droplets of melted ash.

Tilting Furnaces

There are various kinds on rotary kilns, distinguished by their shape (cylindrical, conical), their aspect ratio L/D, or even their rotary movement.

Laurent-Bouillet proposed a particular type of furnace, with a typical conical-cylindrical shape, in which MSW is subjected to an oscillating movement [132].

Multiple Hearth Furnaces (MHF)

Summary MHF have been developed in the nineteenth century for ore roasting and treatment and metallurgical applications are still leading in Europe. Especially in the USA, they have been applied for sewage sludge incineration.

Principles The MHF is a cylindrical construction, composed of a number of circular hearths mounted one above the other. Each hearth contains an air-cooled rabble arm, driven from a common central shaft. Blades, fitted to the slowly rotating rabble arm move the material forward – depending on the angle at which they suspend from the arms – either toward the center or toward the periphery, until it passes over a discharge aperture and falls onto the lower hearth.

The retention time of the charge is varied by changing the speed of rotation of the rabble arms or, rarely, by adapting their relative position to the floor.

Construction and Operation   Multiple hearth furnaces (MHFs) consist of a series of superimposed hearths, solids being fed on top and descending stepwise by gravity, after describing a spiraling movement on each hearth, starting at the discharge point of the higher hearth and ending at that to the lower hearth. Gases generally mount, in countercurrent to the movement of solids, aided by buoyancy. The feed material is charged onto the upper hearth and slowly makes its way down, falling from one hearth to the next, while it is progressively dried, heated, ignited, combusted, and finally cooled by the combustion air. The latter is introduced in part or all at the bottom of the furnace, preheated by the ash on the lower hearth(s), and partly consumed on successive combustion hearths. The resulting flue gas is cooled by the incoming feed and leaves the unit toward possible post-combustion and cleaning. Auxiliary burners are used for preheating and adapting and controlling the temperature profile. The atmosphere is controlled by balanced introduction of air, recirculation, or other means [131].

Applications   MHFs were originally developed for roasting sulfide ores (Nichols-Herreshoff). Later they were adapted for sewage sludge incineration and for competing with fluidized beds. They provide a controllable temperature record to the feed, generally involving sequential drying, heating, reacting, and cooling hearths. There is much contact surface with air and this surface is periodically renewed by the passing rabble arms with attached plates, plowing through the material.

The main application in waste is incinerating sewage sludge and regeneration of spent carbon or lime. The heat required for drying sludge can – when desirable – be supplied by firing pulverized refuse on lower hearths as an auxiliary fuel.

Lucas Furnace Developments, Ltd., once designed a rotary, single hearth furnace, sloping down from the periphery toward the center. It was proposed for incinerating sewage sludge, old tires (without any prior size reduction), and plastics. After preheating the furnace, waste was fed at regular intervals by means of a ram feeder. As the solid hearth slowly rotates the waste first moves along the outer periphery and gradually spirals to the central discharge point. Finally, the ash falls into a quench tank and is removed by a scraper conveyor.

In this Lucas furnace, the gas flow is organized for cyclonic combustion. High-velocity nozzles direct the combustion airflow tangentially into the furnace, cooling the walls to 850–900°C. The central temperature attains 1,450°C. The plant operates at 80–100% excess of air. Operation at reduced capacity suffers from loss of turbulence in the gas phase, a problem that can be tackled using auxiliary steam jets.

Advantages and Disadvantages

MHFs are a proven and traditional technology that allows a flexible adaptation of operating conditions on each hearth. This versatility is less available in rotary kilns or shaft furnaces.

Because of its complex construction this furnace is limited in its maximum capacity. Moreover, it takes long times to preheat and shutdown MHFs, since it is important to avoid thermal shocks.

Fluidized Bed Incinerators

Summary Fluidized (bubbling) bed combustors are exceptionally adaptable, allowing to burn (or gasify, if air supply is sub-stoichiometric) solid, pasty, melting, liquid, sludge, slurried, and gaseous waste, simultaneously and at unusually low temperatures. Moreover, they admirably accomplish in-bed solids mixing and heat transfer and leave a neatly polished solid residue, sinking in the bed. Fine ash is entrained, including particulate formed by attrition or erosion. Desulfurization with limestone or dolomite is possible in bed at combustion temperature and thermal NO_x-formation remains negligible. The principle limitations are the relatively important pressure drop, as well as bed agglomeration, in the presence of tacky ash or salts. Draining decanted residues requires a dedicated circulation circuit of bed materials, adding to mechanical complication: the extracted bed material is sieved and the underflow returns to the bed.

Circulating bed incinerators have been developed and used since the late 1990s by Zhejiang University and also by Tsinghua University. In China, circulating fluid bed units are unusually popular, since combustion stability can be maintained simply by adding cheap coal, instead of using expensive oil [133].

Principles of Fluidization Consider a fixed bed of granular media, such as sand, ash, or limestone, supported on a porous plate, the distributor plate. An upward current of fluid traversing this layer incurs a pressure drop Δp, which rises as the fluid flow rate increases. Meanwhile, the bed porosity (i.e., the void volume) gradually expands. At a given flow rate the pressure drop Δp even equals the pressure, exerted by bed weight. Then, the minimum velocity of fluidization u_{mf} is reached and the head loss corresponds to the weight of the entire bed per unit of cross section (+ the friction loss, omitted here):

$$\Delta p\left(u_{mf}\right) = \rho A H g \qquad (5.1)$$

with Δp = pressure drop, m^2
        u_{mf} = minimum fluidization velocity, $m\ s^{-1}$
        A = bed cross-section, m^2

H = bed height, m

g = acceleration of gravity, m s^{-2}

In principle, the bed thus reaches kind of a state of levitation. However, the fluid (further termed "primary air") will not carry the bed upward, given its granular structure. Rather, excess air trickling through the bed starts forming bubbles at the orifices of the distributor. Such bubbles, after a while, leave the distributor and rise through the bed, more or less like bubbles do in boiling water. The bubbling bed thus resembles a boiling liquid, with series of bubbles rising from the bottom and bursting at the surface. This upward movement of air bubbles creates excellent mixing patterns in the bed leading to temperature homogeneity. Light materials tend to float; dense materials sink to the bottom of the bed. Fluidized beds may be used both as separator of stones or metals present in waste, at low gas velocities, and as a mixer, at velocities a multiple of u_{mf}.

The value of u_{mf} can be estimated using empirical correlations. The gas flows both as bubbles and as a "dense phase" trickling flow. Smooth fluidization is obtained only at a gas velocity of three to four times u_{mf}, together with suitable particle mixing and heat transfer characteristics.

With rising gas velocity, entrainment of fines becomes more important, depending on their terminal falling velocity. The entrainment of bed particles specifies an upper limit of gas velocity. Depending on particle size (often 0.3–0.8 mm), gas velocities typically range between 0.3 and 5 m s^{-1}. There is gradual transition from a bubbling bed to a circulating fluid bed, the latter characterized by a cycle of entrainment, particle separation, and recirculation (Figs. 5.3 and 5.4).

Principles Fluidized bed incinerators burn waste, suspended and moved around erratically in a vigorously bubbling bed of hot granular material. High heat generation rates can be obtained despite the low operating temperatures, typically 750–900°C.

In bubbling fluidized beds, combustion of volatile matter mainly takes place in the freeboard zone, i.e., above the bed. This freeboard zone should be ample and well mixed, so that reducing and oxidizing strands can mix and burn out completely. Secondary air provides the swirl required for mixing. The bed material provides a thermal flywheel that effectively copes with short-term fluctuations in feed rates and quality.

Construction and Operation A fluidized bed incinerator consists (starting below and moving upward) of an empty plenum chamber; a distributor supporting the bed; the bed of granular material to be fluidized; a freeboard zone ensuring disengagement of entrained particles and post-combustion, and adapted feeding; start-up and auxiliary heating burners; waste heat boiler; and pollution abatement equipment. The *distributor* supports the bed material and evenly distributes the primary combustion air over the entire cross section. Even distribution requires the pressure drop over the distributor to be at least 0.1–0.2 times the pressure drop over the bed. The distributor should also prevent weeping of bed material into the

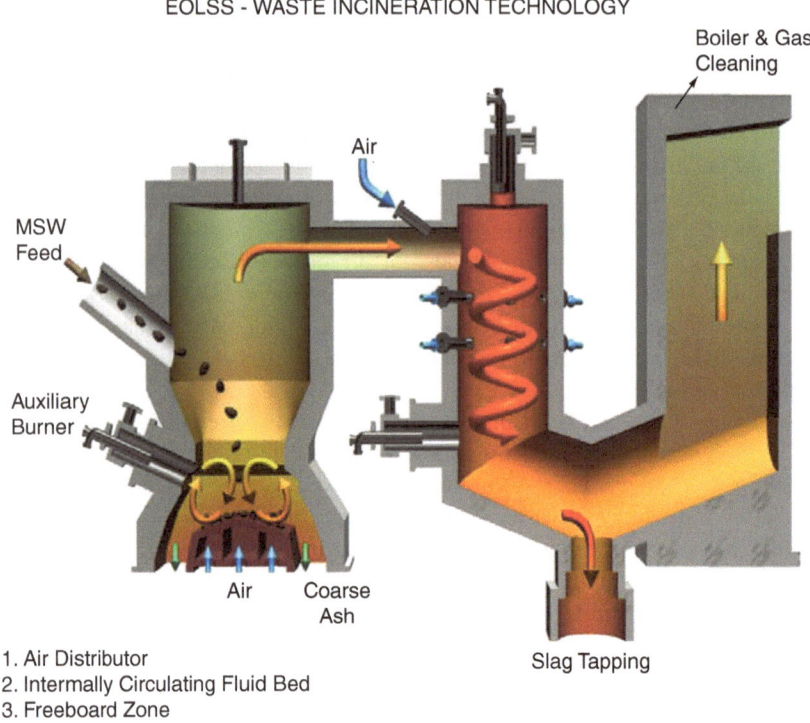

Fig. 5.3 Fluid bed gasifier with slagging post-combustor (By Courtesy of Ebara Co., Japan)

plenum chamber, or its cycling between both zones, with erosion as a consequence. The distributor is made of heat-resistant alloy or forms an arch of refractory material. In principle, there is a wide range of possible designs; a bubble cap distributor is selected most often.

The *bed material* consists of a graded fraction of clean heat-resistant material, such as sand, more seldom alumina, limestone, dolomite, or ash. Pollutants, such as SO_2, can be removed in situ by simply feeding limestone or dolomite into the bed. The bed is preheated to operating temperature by generating hot air in a separate furnace or using a start-up burner directed toward the surface of the bed and then fluidizing gently.

Combustible waste and *auxiliary fuel* are generally fed into the bed to ensure that the heat of combustion is largely generated inside, rather than above the bed, in the freeboard zone. Yet, much of the combustion of the volatile matter will burn above

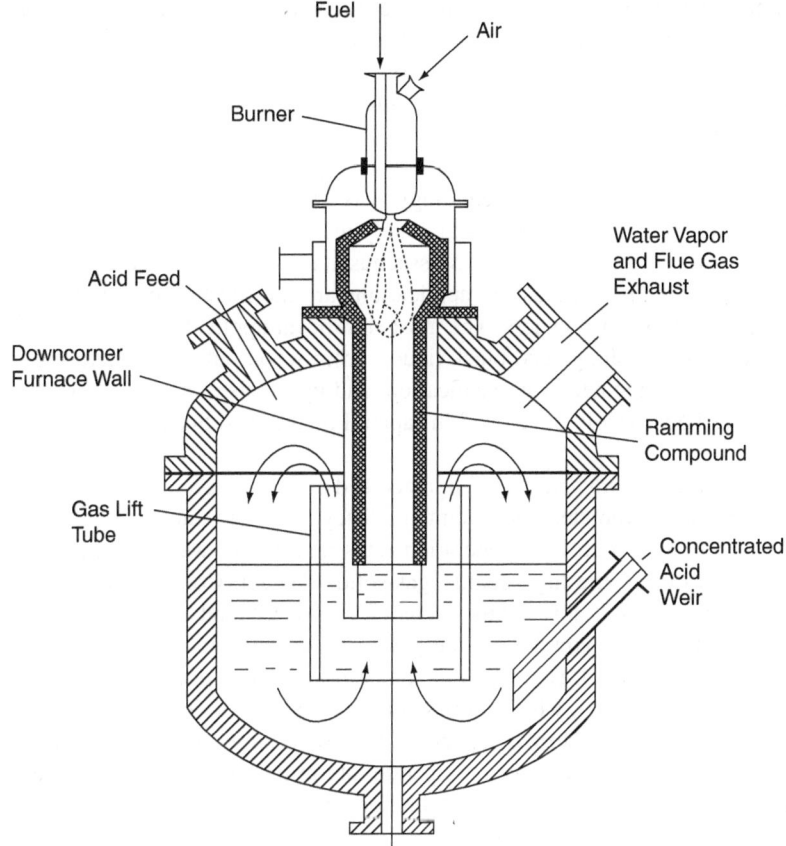

Fig. 5.4 Submerged combustor (Günther [34])

the bed. Gas is fed through independent bubble caps, liquid fuel, slurries and pumpable sludge, and lances, and solids by means of a screw or a pneumatic feeder.

Low-calorific wastes can be dropped onto the top of the bed by a chute fed by a conveyor belt, or sprayed over the bed by means of suitable nozzles. Mechanical spreaders may assist in obtaining a more uniform distribution of the feed. The falling droplets or particles are partly dried while dropping onto the bed. Once in the bed, drying, heating, ignition, and burnout proceed very rapidly. Feeding large pieces lead to the local evolution of excessive amounts of volatiles, causing total depletion of oxygen, evolution of clouds of pyrolysis and gasification products, and, eventually, of sequences of CO and TOC peaks. For this reason the feeding rate should be bit by bit and steady, and feed materials should not be larger than 5 or 10 cm.

Bubbling beds project particles into the freeboard. Most settle in the *freeboard zone*, but the finer ones are partly entrained. Finer particles are separated in internal or subsequent cyclones and flow back into the bed, to complete their combustion. Secondary air is injected into the freeboard zone to complete the combustion.

Applications   Fluidized bed technology was first applied in the 1920s, in coal gasification (Winkler) [134]. Fluid catalytic cracking of gas-oil to gasoline followed during World War II (Massachusetts Institute of Technology, Esso) [134]. Other important industrial applications are the roasting of sulfide ores and the drying of polymer powders. The most significant applications are the incineration of wastewater sludge and black liquors from wood pulp manufacturing [135].

Fluidized bed incineration, gasification, and pyrolysis of shredded or classified refuse have been widely developed in Japan, Finland, and Scandinavia.

Advantages and Disadvantages   Fluidized bed incinerators are relatively simple to build, operate, automate, and maintain. They have no moving parts at high temperatures. Yet, high heat generation rates and bed-to-wall heat transfer rates are obtained due to the high-quality gas-solids contact. Complete combustion is possible already at a low temperature (750–850°C) and a low excess of air (15–35%); hence the volume of flue gas to be cleaned and the NO_x generation rate are relatively small. Bed material can easily be added or removed (draining at the bottom or overflow), which allows adding also lime or dolomite.

Thanks to the large thermal capacity it is also possible to absorb important step changes in feeding rate and even to operate intermittently: cooling of the bed after shutting down is very slow, so that proper operating conditions can rapidly be reached after a standstill of 1 or 2 days.

On the other hand, both the power requirements for fluidization and the dust content of flue gas are quite high. Dense material may segregate and accumulate on the distributor plate, which can be avoided by using an appropriate distributor design, such as sloping distributor plates or arrays of spaced perforated tubes sparging air into the bed. A waste heat boiler and/or preheated air are required to reduce the stack heat losses.

The most serious operating problem occurs when the combustion temperature increases beyond the softening point of the ash. Rapid particle agglomeration then occurs, followed by solidification of part or all of the bed material. When this happens, the solidified material has to be excavated by pneumatic hammers after cooling of the bed.

Vortex Combustors

Summary   Vortex incinerators can be used for high-rate combustion of gaseous, liquid, and finely divided solid fuels or wastes. Larger particles require longer residence times and may not burn out completely; in this case, supplemental mechanical means, such as a specific burnout grate, have to be provided for retaining the burning residue [9].

Principles   Vortex firing involves a highly turbulent mode of combustion, featuring fast transfer of heat and mass and resulting in high volumetric rates of heat release.

Construction and Operation   Fuel (or waste) is blown in tangentially into a conical or cylindrical furnace. The rotary movement of combustibles suspended in combustion air as a carrier creates excellent mixing conditions and hence high combustion intensities and temperature homogeneity.

Two vortices are formed: an outer one consisting of combustion air and waste and an inner one of burning gases. The cooler outer flow shields the refractory walls from overheating and it is rapidly preheated by the hot inner core, considerably stabilizing a steady combustion. A wide range of operating temperatures and a low excess of air can be used. This may lead to a slagging operation.

Applications   Tangential firing involving vortex combustion has been used extensively in coal-fired utility boilers. In one design, pulverized coal is fired in a separate cylindrical vessel, somewhat inclined to the horizontal. In a second design, pulverized coal, together with combustion air, is injected from the four corners of a vertical chamber with a square cross section. Jets are directed tangentially to an imaginary circle contained in this section.

Heat rates in such cyclonic furnaces allow liquid tapping of slag, also with medium HHV waste, such as dry straw or wood chips. Wet bottom cyclonic furnaces for firing coal with an unusually low ash melting point have been developed.

Advantages and Disadvantages   Cyclonic furnaces are compact and highly productive. These advantages weigh more, in case the unit capacity is important. The principle is applied less often for small plants. A high-temperature operation may lead to considerable NOx formation, unless excess oxygen is really minimized.

Slagging Incineration

Summary *Slagging incineration* also termed "wet bottom operation", is an option whenever combustion is conducted at quite high temperatures, or when the resulting ash has an unusually low melting range. This occurs when the waste contains certain groups of chemicals in its ash, such as borates and numerous alkali salts.

Principles *Wet bottom* combustors fire fuel (coal, or waste) at temperatures exceeding the melting point of ash. Incinerators may operate in this slagging mode in case:

- Waste is sufficiently high calorific and combustion temperatures are adequately high
- The resulting slag either has adequate fluidity, or fluxes are added
- Provisions for tapping molten slag are available

Preheating combustion air, enriching it with oxygen, providing auxiliary fuel, and dissipating electric power in the charge (Ohm effect, electric arc, hot plasma) all allow to raise combustion temperature to higher levels and addition of fluxing agents (fluorspar, iron oxides, lime) may be used to lower the melting range and enhance slag fluidity. Molten slag can be tapped discontinuously and discharged to solidify to large crystalline blocks. Generally, however, it is quenched by pouring the melt into a water bath, converting it into small glassy grains. Continuous tapping is uncommon, given the small capacity. When treating metal rich waste, two phases might be formed: a light slag floating on top of molten metal.

Construction and Operation There are numerous different methods to conduct incineration or even gasification under slagging conditions. Such methods encompass, e.g.:

- Shaft furnaces, operating like a blast furnace. Examples: Lurgi pressurized moving bed oxygen/steam gasifiers. Union Carbide, Andco-Torrax and Nippon Steel gasifiers.
- Rotary kiln incinerators operating on high-calorific waste and in a slagging mode.
- Mechanical grate incinerators, fitted with a dedicated furnace to melt the residue.
- Electric arc furnaces (cfr. metallurgical Industry).
- Suspension firing. Cfr. The Vortex Furnace and Koppers-Totzek Coal Gasifiers.

Applications Already in 1934 Rummel [134] pioneered the concept of using molten material (these could be slag, metals, or salts) as heat carrier and oxygen transfer agent. Ever since these 1930s, slagging incinerators have been experimented with, in association with cokes addition, electric arc furnaces, plasma torches, with special waste with unusually low ash melting traject, or with waste streams, warranting high disposal cost, e.g., radioactive waste, or PCBs.

Recently, slagging operation in Japan became a standard, following the necessity of converting incinerator residues into glassy slag.

Nippon Steel has developed blast furnace technology, applied to MSW.

Ebara Co. has pioneered a fluidized bed gasification plant, featuring postcombustion under slagging conditions of the gas.

Advantages and Disadvantages   Slagging incineration has several potential advantages, such as

- Simpler furnaces, ash flowing along an inclined floor
- Generating dense glassy slag, with low leaching rates and almost free from combustible inclusions
- Operating at low excess of air, reducing flue gas flows to be cleaned as well as stack losses.

The major problem is ensuring steady fluidity of slag, while still limiting its attacks on refractory. A thin layer of solidified slag may be maintained on the refractory lining to cover and protect the refractory (Fig. 5.5).

It is recommendable to separate combustion from ash melting, by providing a controlled supply of heat and possibly flux in the slag tapping area. Slagging operation exists in many variants. Heat supply is secured by the following means, alone or in combination:

- Plasma torch
- Addition of coke
- Auxiliary burners
- Combustion of gasification products

Example: Incineration in a Molten Salt Bath   Incineration in a molten salt bath has also been applied when dealing with hazardous wastes, such as pesticides, explosives, etc. Combustion in a bath of molten salts (e.g., sodium carbonate, potassium carbonate,) could present the following advantages:

- Molten salt acts as a heat carrier and combustion catalyst, ensuring complete oxidation at temperatures lower than normally required.
- Carbonized residues and dust and ash particles are entrapped in the bath.
- Acidic pollutants in the off-gas react with the salt and are also retained in the bath.

A potential disadvantage is the required disposal or regeneration of spent salt. Moreover, managing volatilizing salts is problematic, since deposits of desublimated salt fumes will need to be removed periodically.

Submerged Combustion

Summary   Submerged combustion combines a vertical conventional combustion chamber with immediate direct-contact quenching of flue gas by its immersion in aqueous liquor, brine, lye or acid, ensuring fast heat and mass transfer in a bubbling liquid mass.

Principles   Submerged combustors feature a vertical, refractory lined steel shell furnace equipped at the top with a down-firing burner, the flue gas of which bubbles through a reservoir filled with quenching and scrubbing water or other aqueous liquids (Fig. 5.6).

Construction and Operation   The plant consists of a slender, elongated, vertical furnace, plunging as a downcomer tube into a bath retained in a wider container. Frequently, a concentric tube surrounds the downcomer, forming an annular space, acting as an airlift and promoting internal mixing of the liquor contained in a quenching bath. As a result the gas is suddenly quenched, freezing undesirable reactions, such as forming chlorine according to the Deacon equilibrium of reaction (13).

Waste sulfuric acid, brine, or other corrosive solutions can thus be concentrated by direct contact between hot flue gas and the liquor to be treated. Heat and mass transfer between flue gas and liquid quench are almost instantaneous: the quenched flue gas is almost completely saturated with water vapor and leaves substantially at the temperature of the quenching bath.

Applications   Submerged combustors have a long tradition, with first patents in the nineteenth century. Chemico used it as a means to concentrate dilute sulfuric acid, Nittetu Chemical Engineering to treat chlorinated waste [137]. It is mentioned in the IPPC report [136]. Submerged combustors are used to:

- Quench and clean flue gas, arising from the combustion of chlorinated organics [137] or CFCs [138, 139]
- Concentrate wastewaters or corrosive acids, maintained as quenching bath
- Recover a solution of inorganic salts, when firing aqueous solutions of organic or inorganic salts
- Recover dilute hydrochloric acid of acceptable concentration when firing chlorinated wastes

The liquid wastes are atomized and injected into the flame, so that they are rapidly dried, thermally decomposed, and completely combusted. Inorganic compounds are converted to tiny molten salt particles and are recovered as a salt solution or slurry in the quench vessel.

Another design (Nittetu) features a fractionating column mounted on top of the water vessel. Wastewater contaminated with volatile hydrocarbons is fed on top of the column and hydrocarbons are stripped off in contact with a rising mixture of flue gas and water vapor. After condensation of the vapors, the condensed heavier hydrocarbons are recycled to the quench vessel. The noncondensable is combusted.

Advantages and Disadvantages   Submerged combustors are relatively simple, efficient equipment. Quenching and scrubbing are fast and direct-contact. Chlorine formation is suppressed when firing halogenated solvents and vapors. Also dioxins formation is suppressed.

A dedicated website is [140]; various hardware are presented in [141] (Table 5.1).

Chapter 6
Refuse-Derived Fuel

Rather than firing waste as it comes, one can convert it into storable fuel, following a suitable sequence of operations, composed of primary and secondary shredding, grading, wind sifting and screening, magnetic and eddy-current separation, etc. Suitable combinations of such operations may convert municipal solid waste, packaging, wood, paper and plastics, etc., into better manageable and storable refuse-derived fuel (RDF) with more predictable characteristics and specifications, such as HHV, and proximate and ultimate analysis. RDF assumes different forms, such as fluff, powdered (after adding embrittling agents), or densified, i.e., in bales, pellets. Already in the 1970s, the National Centre for Resource Recovery (Washington) tried to standardize RDF, to improve its acceptance and access to the energy markets [9, 142].

The preparation of RDF may proceed according to very simple as well as more complex schemes, promising higher quality as well as more investment and operating cost. EcoFuel® was a powdered product, obtained by raising the temperature and adding an embrittling agent, e.g., sulfuric acid. In one case, processing started by wet pulping (Black Clawson at Franklin, Ohio); the resulting RDF was wet during processing, which is counterproductive for thermal utilization. Moreover, the only large plant in Florida suffered from odor problems and was eventually dismantled. Yet the processing generates clean fractions of metals and glass.

Today, mechanical biological treatment [143, 144] is proposed as a generic group of processes to produce a fuel fraction and sorting fractions.

The complete combustion of solids generally requires residence times of typically half an hour, except for some high-intensity incinerators, such as those firing a *refuse-derived fuel* (RDF). In principle, the residence time available for combustion is comparable now to that of flue gas, i.e., a matter of seconds only for

This chapter was originally published as part of the Encyclopedia of Sustainability Science and Technology edited by Robert A. Meyers. doi:10.1007/978-1-4419-0851-3

A. Buekens, *Incineration Technologies*, SpringerBriefs in Applied
Sciences and Technology, DOI 10.1007/978-1-4614-5752-7_6,
© Springer Science+Business Media New York 2013

suspension firing up to about a minute in bubbling fluidized bed plants. Circulating fluidized bed units feature cyclonic separators that collect coarse matter for recycling. In practice, the denser RDF will thus be retained by suitable aerodynamic and geometrical means (gravity or centrifugal force) that extend its residence time until the residue is so fine that it is entrained.

RDF preparation is expensive, typically in the range of 20–80 US$/Mg. These costs derive both from investment and operations and include power consumption and heavy wear on equipment. Fire and explosion hazards are notoriously present during shredding, drying, and even longer-term storage.

Co-firing of Waste or of RDF

In most cases waste is incinerated in dedicated furnaces. In some instances, it may be more attractive to incinerate waste in preexisting plants, such as industrial furnaces, power plant, cement kilns, etc. The advantages are obvious:

- Investment cost is limited to the additional plant, required to prepare, store, feed, and fire the waste.
- The energy content of waste is put to good use, often at much higher efficiency than is possible in a dedicated plant. A thermal power plant typically operates at an efficiency of HHV to power of ca. 44%, against typically 16–24% for dedicated incinerators.

Theoretically, it would thus be attractive to replace dedicated incineration by usage of waste as a fuel. Yet there also disadvantages, such as below:

- Integrating distinct activities (waste elimination and heat and power generation) also means declining the operating flexibility of each individual activity.
- Incineration is subject to much more stringent emission codes, compared to those for thermal power plants, industrial furnaces, or cement kilns. Co-firing has been denigratingly termed "solution by dilution." The EU directive on incineration has considered this problem and offered a solution featuring flexible emission limits.
- The waste composition should be confronted with that of the conventional, generally solid, fuel with respect to the pollutants S, N, Cl, and heavy metals.
- The ash arising from waste may affect and often lower product quality, e.g., in cement and especially limekilns.
- Some elements contribute not only to pollution, but also to the creation of operating problems,such as superheater fouling and corrosion (biomass or RDF co-firing) or cycling of heavy metals (cfr. Cement Kilns).

Table 6.1 Typical combustion conditions

Combustion conditions	Combustion temperature (°C)	Furnace temperature (°C)	Gas velocity (m s⁻¹)	Residence time (s)	Air number (–)	Thermal volumetric load (MW m⁻³)	Thermal cross-sectional load (MW m⁻²)
Power plant:							
Natural Gas	1,100–1,400	1,000–1,100	5–10	1–3	1.05–1.1	0.25–0.35	5–8
Oil	1,100–1,400	1,000–1,100	5–10	1–3	1.05–1.2	0.25–0.35	5–8
Grate stoker	1,100–1,300	1,000–1,100	4–9	1–3	1.3–2.5	0.15–0.35	0.5–2.5
Fluidized bubbling bed combustor	750–1,050	750–1,050	0.5–5	1–3	1.2–1.4	2–5	1–2
Circulating fluid bed combustor	750–950	750–950	5–8	0.5–6	1.12–1.3	8–20	2–8
Pulverized coal firing	1,100–1,500	1,050–1,250	5–10	1–3	1.13–1.3	0.06–0.3	0.6–3
Pulverized lignite firing	1,100–1,300	950–1,150	4–8	1–3	1.2–1.5	0.06–0.15	2.5–5
Pulverized coal firing (wet bottom)	1,300–1,600	1,000–1,150	5–10	1–3	1.15–1.3	0.1–0.4	4–6
Mechanical grate MSW-incinerator	1,000–1,250	1,000–1,100	3–8	3–6	1.5–2.0	0.15–0.35	1.4–1.6
Fluid bed sludge combustor	750–900				1.05–1.8	1–3	0.5–1
Stationary combustion chamber					1.2–3	0.1–0.3	0.1–1
Rotary kiln incinerator					1.6–3.5	0.15–0.2	1.5–2.5
Postcombustion chamber	1,050–1,300	1,050–1,250			>1.4	0.08–0.35	1–1.5

Source: Compiled from Görner

Thermal Power Plants

Co-firing RDF in thermal power plants offers solutions that hold the promise of limited investment, related to the production (possibly off-site), storage, and firing of RDF. Conversely, RDF co-firing may create serious problems at the level of boiler fouling and emissions.

Thermal power plants in general are large-capacity units (40–400 MW_{el}), typically one order of magnitude larger than the usual waste-to-energy (WtE) projects. As they stand, they are fully equipped with provisions for fuel supply, firing, and ash storage, boiler feedwater production, steam generation at high pressure, turbo-alternator, transmission lines, steam cooling, and condensation provisions. Sharing these provisions with incineration plant allows sharing all provisions related to the steam circuit and power generation.

Co-firing of RDF has been proposed consistently since the early 1970s. Ideally, the hosting power plant fires solid fuels such as coal or lignite. The RDF must be reduced in size, so that the individual particles burn out in suspension. Dense parts falling out can be collected on a dump grate for completing their combustion.

Co-firing of biomass has also been considered, at first in Denmark, to eliminate the polluting practice of field burning. Biomass is often lean in pollutants (sulfur, nitrogen, heavy metals). Unfortunately, the ash is also rich in low-melting potassium salts and hence tacky, causing extensive superheater fouling and corrosion. Pure wood is low in ash (0.5–2 wt.%), yet real biomass, such as straw, is much higher, up to 8 wt.%.

Cement and Lime Kilns

Cement (and lime) kilns are increasingly used for incinerating hazardous and also high-calorific waste. The kilns always operate in countercurrent (Fig. 6.1) and feature combustion temperatures of almost 2,000°C, with kiln lengths ranging from some 50 m (dry process) to about 150 m (wet process). This ensures longer residence times at temperatures above 850°C than any other furnace. Even hazardous pollutants, such as PCBs, requiring a destruction efficiency of at least six 9s (99.9999%), are completely combusted in such kilns. The waste is fired at the lower end of the kiln so that all flue gas starts at flame temperature and then remains in contact with the high temperature reaction zone in which the clinker is formed. Most ash drops out at high temperature and is incorporated into the clinker.

Wet kilns are even longer, since the raw materials mix is to be dried, dehydrated, decarbonated (conversion of $CaCO_3$ into lime), and eventually reacted to clinker at around 1,500°C.

Wet kilns allow separating dust from flue gas and sluicing it out, since the feed enters as a paste. Much shorter dry kilns do not have this feature, the incoming dry meal being heated in direct contact in batteries of cyclonic heat exchangers and

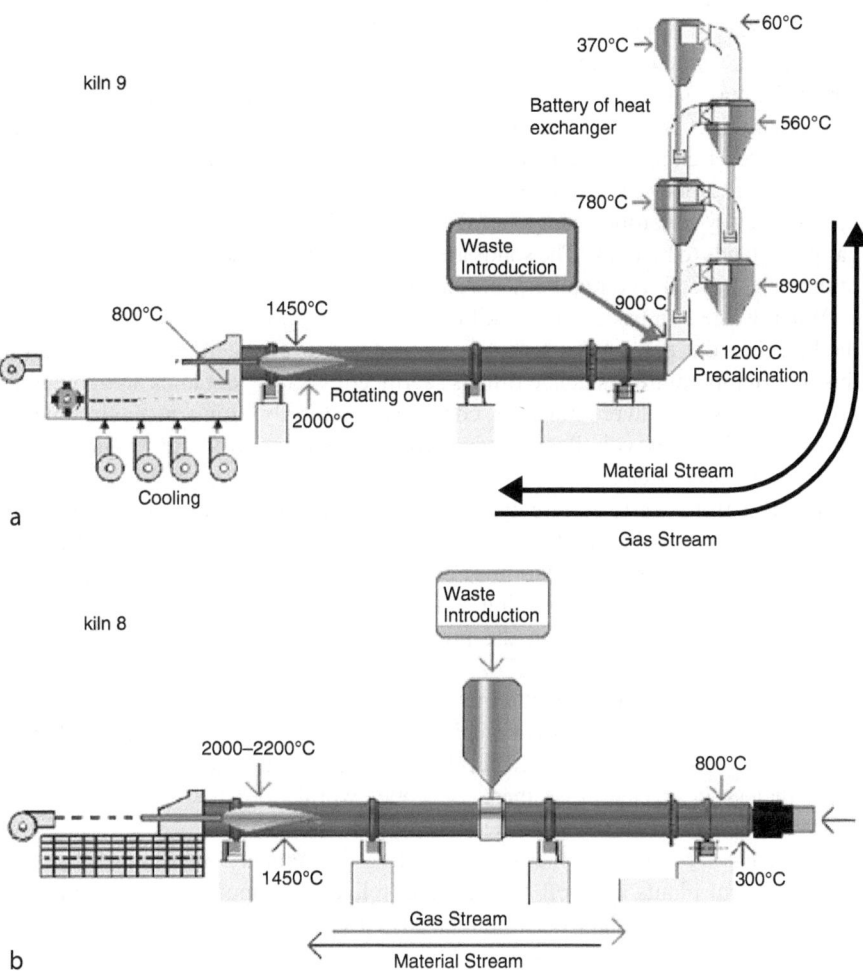

EOLSS - POLLUTION CONTROL THROUGH EFFICIENT
COMBUSTION TECHNOLOGY

Fig. 6.1 Cement kilns treating contaminated soil with feed point (**a**) after the cyclonic heat exchangers and (**b**) mid-kiln. (14**a**) Shows a battery of four cyclonic heat exchangers, featuring direct contact with hot rising flue gas. The feed entering the plant is dried and heated from 60°C to 900°C, completely calcining the limestone in the feed. In the rotary kiln, it converts into clinker. In this unit, contaminated soil is added at the kiln entrance and it is not certain how far emerging volatiles are still combusted completely. (14**b**) Shows an alternative with mid-kiln feeding, and still converts the contaminated soil into clinker, yet leaves more room for post-combustion than in the first case

enter the much shorter kiln already at high temperature, after preliminary de-carbonatation at ca. 800°C. Because of energetic considerations, dry kilns are always preferred for cement manufacturing.

Waste serves primarily as substitute fuel. Yet, waste with appropriate mineral composition (silica, alumina, lime, and iron) may also replace natural feed materials (clay, shale, limestone), alleviating the needs for quarrying (Fig. 6.2).

Cement kilns are thus prime substitutes for hazardous waste incinerators, as long as the wastes are introduced as solid or liquid fuels at the lower, kiln discharge side. There is sufficient oxygen, temperature, and time to complete combustion of even the most refractory hazardous compounds, such as PCBs, even though the mixing in the gas phase tends to be weak. The solids are in better contact, due to the tumbling action of the kiln. Typical feed requirements are presented in Table 6.

When waste is introduced mid-kiln, however, or – worse – at the higher end of the kiln (in a dry plant, yet after the battery of heat exchangers), a sizeable part of this high-temperature residence time is sacrificed. Feeding organics along with the raw materials, however, must be considered carefully from an environmental viewpoint, since any volatiles evolving from the feed report to the off-gas without post-combustion or cleaning.

The ash from waste is largely incorporated into the clinker. This has raised questions regarding the eventual leaching of any heavy metals from clinker, as well as regarding the state of oxidation of chromium, i.e., Cr^{III} or Cr^{VI}. Halogens and volatile heavy metals create cycling and emission problems. Several heavy metals (Pb, Zn, and Cu) volatilize in the presence of chlorides, yet de-sublimate and deposit during cooling.

Cement kilns are important emitters of carbon dioxide, dust, and nitrogen oxides. Emission limit values are much more lenient than for dedicated incinerator plants. The cement route has hence been denigrated as "solution by dilution." Nevertheless it is clear that there is scope, worldwide, for the cement route especially in countries devoid of dedicated plant.

There is very extensive literature regarding the cement route [145–148]. Individual cement plants are well documented, relative to inputs and emissions. The lime route has been much less publicized. Obviously, any ash reporting to the product deteriorates product quality.

Chapter 7
Public Image of Incineration

Incineration has been branded as a substantial source of environmental pollution (dioxins, heavy metals), as well as an easy way around voluntary or even mandatory recycling. Greenpeace has been quite vocal in these criticisms, attacking incineration, and also PVC as a source of dioxins [149–152]. In the meantime it became clear that dioxins in MSW incineration are formed at comparable rates whether or not PVC is present: the element chlorine is so ubiquitous that its concentration in MSW as a rule is not rate controlling [153]. Much more important factors are the steady quality of combustion, the catalytic effects of transition metals, and dioxins forming increasingly in electrostatic precipitators as their operating temperatures rise (Fig. 7.1).

Plume dispersion computations show that immission values of state-of-the-art incinerators are negligible, compared to background values. Such deposition values also have been measured many times [154]. Epidemiological studies relating incinerator emissions to public health never clearly condemned the old generations of incinerators, let alone current units with much lower emission values [155, 156]. One reason for these results is that the body burden of most toxic organics relates to food uptake, rather than to inhalation [157]. Today, the main medical interest is related to minute submicron particles that readily migrate through the lung membranes. Also prenatal exposure to dioxins and POPs has been studied.

Other important aspects are related to absorption, digestion, and eventual health effects of dioxins and dioxin-like compounds, in particular polychlorinated diphenyls (PCBs). PCBs are man-made chemicals, yet they also form (with a different congener profile) in thermal processes, typically representing less than 5% of dioxins and furans, when expressed in toxicity equivalents (TEQs). Dioxin diets were established in numerous countries, since food is the major route to take up TEQs. One pathway between the incinerator stack and the food chain is as follows: emission – particle deposition – absorption by grazing cattle – secretion with the milk. In the past, cow milk has been declared unfit for consumption around both incinerator and metallurgical plants. Halting the emissions at Zaandam (the Netherlands) rapidly restored milk quality. Far higher dioxin emission levels at Gien (France) produced no undesirable effects: dioxins were emitted in the gas

This chapter was originally published as part of the Encyclopedia of Sustainability Science and Technology edited by Robert A. Meyers. doi:10.1007/978-1-4419-0851-3

A. Buekens, *Incineration Technologies*, SpringerBriefs in Applied Sciences and Technology, DOI 10.1007/978-1-4614-5752-7_7,
© Springer Science+Business Media New York 2013

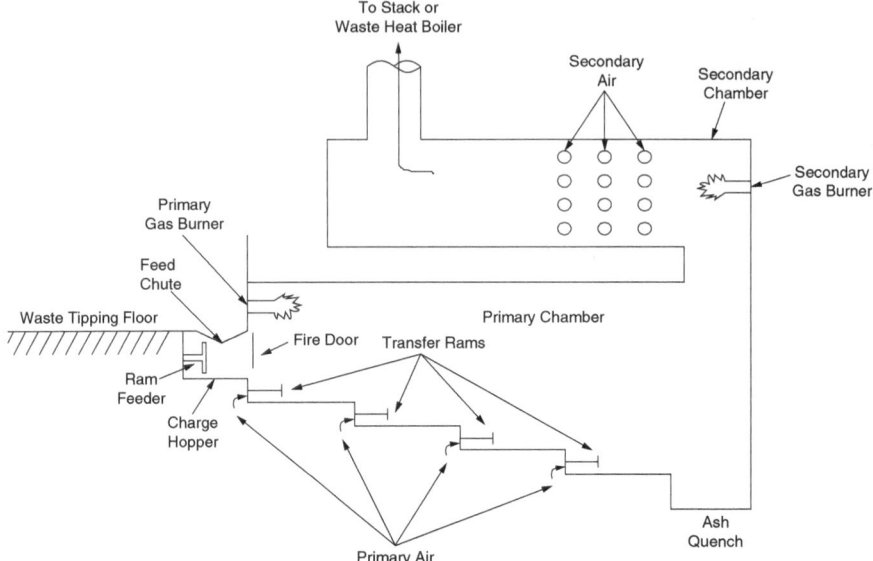

Fig. 7.1 Starved air incinerator (By courtesy of)

phase and apparently degraded in the atmosphere, rather than impairing the quality of dairy products.

Incinerators are no longer major polluters, given the extremely stringent emission codes applied today. There is a tendency at present to compare incineration to its traditional alternatives, landfill and composting, incorporating additional criteria, concepts and methods, such as the impact upon climate change and global warming. Landfill is responsible for important greenhouse gas emissions; evolving fermentation gas contains carbon dioxide and also methane, a much more potent greenhouse gas. Composting also emits greenhouse gases, yet does not get the bonus of producing green energy. Markets for compost are as precarious as those for incinerator heat.

Waste incineration is also held responsible for destroying values, available for recycling. Yet, the main bottlenecks of recycling are markets for secondary raw materials showing low-grade specifications or containing *pernicious contraries*: any imbalance between supply and demand exerts strong leverage on market prices.

Chapter 8
Future Directions

At present, *waste incineration* has evolved to mature technology, with mechanical grate incinerators as standard in MSW incineration, rotary kiln plant for firing industrial and hazardous waste, and fluidized bed units for sewage sludge, as well as for co-firing wastes with extremely dissimilar properties. Yet, each of these units still has some technical limitations. Mechanical grate stokers must support the waste during combustion, yet have difficulty in coping with both very wet and high-calorific waste. For rotary kiln units, gas phase mixing and wear are major problem areas. Fluid bed units require steady, size-reduced feed and may experience loss of fluidization in the presence of low-melting ash and at too high temperatures.

Flue gas cleaning has evolved considerably since the early 1970s, under pressure of ever tightening limit values (Tables 8.1 and 8.2). Given the thorough cleaning generally practiced, it seems unlikely that these limit values would evolve even further. In particular cases, e.g., dioxins, emission values were promulgated on the basis of rather thin evidence and in the absence of proven technology to reach the new limit values. Yet, it cannot be excluded that still new parameters would be brought forward, such as nanoparticles and nitrous oxide (N_2O). However, it is obvious that more can be gained by cracking down on open burning of waste and other primitive and polluting forms of combustion.

The concept of *refuse-derived fuel* (RDF) production holds the promise of conducting incineration in non-dedicated units, such as thermal power plant, and cement and lime kilns. Although RDF is still produced and fired in such plants, initial promise has been mitigated by the added cost and complexity of fuel preparation, by both environmental and product quality (lime, cement) concerns, and by logistic and operational requirements.

This chapter was originally published as part of the Encyclopedia of Sustainability Science and Technology edited by Robert A. Meyers. doi:10.1007/978-1-4419-0851-3

A. Buekens, *Incineration Technologies*, SpringerBriefs in Applied
Sciences and Technology, DOI 10.1007/978-1-4614-5752-7_8,
© Springer Science+Business Media New York 2013

Gasification and *pyrolysis* processes have frequently been proposed and tested at laboratory, pilot, and full-scale level. Their theoretical advantages, such as simpler operation and lower volumetric rates of gas production, have materialized in practice in only few cases. A decisive disadvantage is poor reliability and availability. A large majority of actually constructed plants have actually been scrapped after realizing precarious operating records (e.g., Siemens at Fürth, Thermoselect at Karlsruhe). Some communities were forced to pay for an entire generation for such plants (Andco-Torrax plant in Grasse). By far the most experience has been gathered in Japan, with slagging shaft furnace operation (Nippon Steel) and fluidized bed gasification, followed by post-combustion and fly ash melting and granulating (Ebara Co.) as most successful representatives.

Slagging operation produces glassy aggregate, rather than clinker and fly ash. The question rises whether the more attractive residue can justify considerable supplemental cost, higher energy consumption, and lower availability.

Some organizations important in matters of incineration:

- Air & Waste Management Association (A&WMA)
- American Academy of Environmental Engineers (AAEE)
- American Institute of Aeronautics & Astronautics (AIAA)
- American Institute of Chemical Engineers (AIChE)
- American Society of Mechanical Engineers (ASME)
- Chartered Institution of Wastes Management, London
- Coalition for Responsible Waste Incineration (CRWI)
- Electric Power Research Institute
- Institute for Professional Environmental Practice (IPEP)
- Institute of Chemical Engineers – United Kingdom (IChemE)
- Institution of Mechanical Engineers – United Kingdom (IMechE)
- Integrated Waste Services Association
- International Solid Waste Association (ISWA)
- Japan Waste Management Association ()
- Korea Associate Council of Incineration Technology (KACIT)
- Korea Society of Waste Management (KSWM)
- National Institute for Environmental Studies ()
- National Institute of Environmental Health Sciences (NIEHS)
- Society of Chemical Engineers – Japan
- Solid Waste Association of North America
- Swedish Chemical Society – Sweden
- UK Environmental Agency – United Kingdom
- United States Department of Energy (US DOE)
- United States Environmental Protection Agency (US EPA)
- Waste-to-Energy Research and Technology Council (WtERT)

Bibliography

Primary Literature

1. Lewis H (2007) Centenary history of waste and waste managers in London and South East England, Chartered Institution of Wastes Management, London. http://www.iwm.co.uk/web/FILES/LondonandSouthernCentre/London_and_Southern_Centenary_Histroy.pdf. Accessed July 2011
2. Kleis H, Dalager S (2007) 100 years of waste incineration in Denmark – from refuse destruction plants to high-technology energy works. DTU, Copenhagen. http://www.ramboll.com/services/energy%20and%20climate//media/Files/RGR/Documents/waste%20to%20energy/100YearsLowRes.ashx. Accessed July 2011
3. Reimann DO (1991) Abfallentsorgung mit integrierter Abfallverbrennung – Verfahren von gestern und heute. In: Reimann DO (ed) Rostfeuerungen zur Abfallverbrennung. EF-Verlag für Energie und Umwelt, Berlin, pp 1–20
4. Reimann DO (1991) Die Entwicklung der Rostfeuerungstechnik für die Abfallverbrennung – Vom Zellenofen zur vollautomatischen, emissions- und leistungsgeregelten Rostfeuerung. In: Reimann DO (ed) Rostfeuerungen zur Abfallverbrennung. EF-Verlag für Energie und Umwelt, Berlin, pp 21–60
5. Reimann DO (1991) Rostfeuerungen zur Abfallverbrennung. EF-Verlag für Energie und Umwelt, Berlin
6. Tanner R (1965) Die Entwicklung der Von Roll-Müllverbrennungsanlagen. Schweizer Bauzeitung 83(16)
7. Picture of the first Hamburg incinerator (1985) http://fr.wikipedia.org/wiki/Fichier:Erste_M%C3%BCllverbrennungsanlage_Hamburg.jpeg. Accessed 29 Dec 2011
8. Schoeters J (1975) Patent study on mechanical grate development. VUB, Brussels
9. Buekens A, Schoeters J (1984) Final Report Thermal methods in waste disposal – pyrolysis, gasification – incineration – RDF-firing, Contract Number ECI 1011/B 7210/83B
10. Ebara Co. (1993) Fluidised-bed combustion of municipal solid waste in Japan. Company document
11. Buekens A (1978) Resource recovery and waste treatment in Japan. Resour Recov Conserv 3(3):275–306
12. Buekens A (2008) Schmelzverfahren – erfahrungen in Japan. In: Bilitewski B, Urban AI, Faulstich M (eds) Schriftenreihe des Fachgebietes. Abfalltechnik Universität, Kassel
13. Global Environment Centre Foundation, Japanese Advanced Environment Equipment, http://www.gec.jp/JSIM_DATA/company_index.html
14. E.U. (2009) E.U. Guideline for safe and eco-friendly biomass gasification (gasification – guide). http://www.gasification-guide.eu/. Accessed 11 July 2011

A. Buekens, *Incineration Technologies*, SpringerBriefs in Applied Sciences and Technology, DOI 10.1007/978-1-4614-5752-7,
© Springer Science+Business Media New York 2013

15. Buekens A, Bridgwater AV, Ferrero GL, Maniatis K (eds) (1990) Commercial and marketing aspects of gasifiers. Commission of the European Communities, Elsevier Applied Sciences, Luxembourg, pp 1–239
16. Malkow T (2004) Novel and innovative pyrolysis and gasification technologies for energy efficient and environmentally sound MSW disposal. Waste Manag 24(1):53–79
17. Buekens A, Masson H (1980) Wood waste gasification as a source of energy. Conserv Recycl 3(3–4):275–284
18. Siemons RV (2002) A development perspective for biomass-fuelled electricity generating technologies. PhD thesis, University of Amsterdam. http://www.cleanfuels.nl/Projects% 20&%20publications/Siemons_PhD%20Thesis_Internet.pdf. Accessed 11 July 2011
19. Scheirs J, Kaminsky W (2006) Feedstock recycling and pyrolysis of waste plastics. Wiley, Chichester
20. Inguanzoa M, Dominguez A, Menéndez JA, Blancoa CG, Pisa JJ (2002) On the pyrolysis of sewage sludge: the influence of pyrolysis conditions on solid, liquid and gas fractions. J Analy Appl Pyrol 63(1):209–222
21. Buekens A, Schoeters J (1980) Basic principles of waste pyrolysis and review of European processes. ACS Symposium Series 130:397–421
22. Buekens A (1978) Schlussfolgerungen hinsichtlich der praktischen Anwendung der Hausmüllpyrolyse aufgrund weltweiter Erfahrungen. Müll und Abfall 12(6):184–191
23. 12th international congress on combustion by-products and their health effects: combustion engineering and global health in the 21st century – issues and Challenges, Zhejiang University in Hangzhou, China, 5–8 June 2011
24. Chandler AJ, Eighmy TT, Hartlén J, Hjelmar O, Kosson DS, Sawell SE, van der Sloot HA, Vehlow J (1997) Municipal solid waste incinerator residues. Elsevier, Amsterdam\Lausanne \New York\Oxford\Shannon\Tokyo
25. Izquierdo M, López-Soler A, Ramonich EV, Barra M, Querol X 2002) Characterisation of bottom ash from municipal solid waste incineration in Catalonia. J Chem Technol Biotechnol 77(5):576–583
26. Vehlow J (2002) Bottom ash and APC residue management. Expert meeting on power production from waste and biomass – IV, Hanasaari Cultural Center, Espoo, 8–10 Apr 2002. VTT Information Service, Espoo, pp 151–176
27. Sakai S, Hiraoka M (2000) Municipal solid waste incinerator residue recycling by thermal processes. Waste Manag 20:249–258
28. Bergfeldt B, Jay K, Seifert H, Vehlow J, Christensen TH, Baun DL, Mogensen EPB (2004) Thermal treatment of stabilized air pollution control residues in a waste incinerator pilot plant. Part 1: fate of elements and dioxins. Waste Manag Res 22:49–57
29. Baun DL, Christensen TH, Bergfeldt B, Vehlow J, Mogensen EPB (2004) Thermal treatment of stabilized air pollution control residues in a waste incinerator pilot plant. Part 2: leaching characteristics of bottom ashes. Waste Manag Res 22:58–68
30. Achternbosch M, Richers U (2002) Materials flows and investment costs of flue gas cleaning systems of municipal solid waste incinerators. Forschungszentrum Karlsruhe Wissenschaftliche Berichte (FZKA), Karlsruhe, 6726
31. CBR (2011) Personal communication
32. ARGUS – ARBEITSGRUPPE UMWELTSTATISTIK (1981) Bundesweite Hausmüllanalyse 1979/80. Umweltbundesamt, Berlin. Forschungsbericht 103 03 503.
33. ARGUS –ARBEITSGRUPPE UMWELTSTATISTIK (1986) Bundesweite Hausmüllanalyse 1983-1985;Laufende Aktualisierung des Datenmaterials. Umweltbundesamt, Berlin. Forschungsbericht 103 03 508
34. Görner K (1991) Technische verbrennungssysteme, grundlagen, modellbildung, simulation. Springer, Berlin\Heidelberg\New York, p 27
35. Niessen WR (2010) Combustion and incineration processes: applications in environmental engineering. Taylor and Francis, Baco Raton
36. Brunner CR (1996) Incineration systems handbook. Incinerator Consultants, Reston

37. Hämmerli H (1991) Grundlagen zur Berechnung von Rostfeuerungen. In: Reimann D (ed) Rostfeuerungen zur Abfallverbrennung. EF-Verlag, Hrsg

38. European Commission (2006) Integrated pollution prevention and control – reference document on the best available techniques for waste incineration

39. http://en.wikipedia.org/wiki/File:Et_baal.jpg

40. Wilkes JW, Summers CE, Daniels CA, Berard MT (2005) PVC handbook. Hanser Verlag, MŘnchen

41. Buekens A (2006) Introduction to feedstock recycling of plastics. In: Scheirs J, Kaminsky W (eds) Feedstock recycling and pyrolysis of waste plastics: Converting waste plastics into diesel and other fuels. John Wiley & Sons

42. Buekens A (2008) Solving emission problems in a fluid bed MSWI. In: 5th i-CIPEC: international conference on combustion, incineration/pyrolysis and emission control – eco-conversion of biomass and waste, Chiang Mai

43. Briner E, Roth P (1948) Recherches sur l'hydrolyse par la vapeur d'eau de chlorures alcalins seuls ou additionnés de divers adjuvants, Helv Chim Acta 31(2):1352–1360

44. Buekens A, Schoeters J (1986) PVC and waste incineration. APME, Brussels

45. Chimenos JM, Segarra M, Fernández MA, Espiell F (1999) Characterization of the bottom ash in municipal solid waste incinerator. J Hazard Mater 64(3):211–222

46. Meima JA, Comans RNJ (1997) Geochemical modeling of weathering reactions in municipal solid waste incinerator bottom ash. Environ Sci Technol 31(5):1269–1276

47. Commission Decision of 3 May 2000 replacing Decision 94/3/EC establishing a list of wastes pursuant to Article 1(a) of Council Directive 75/442/EEC on waste and Council Decision 94/904/EC establishing a list of hazardous waste pursuant to Article 1(4) of Council Directive 91/689/EEC on hazardous waste (notified under document number C (2000) 1147)

48. Wikipedia, Hazardous Waste

49. Buekens A (2011) Hazardous waste and pollution prevention, course organized by VMAC, Premier Provider of Business Intelligence, Abu Dhabi (U.A.E.)

50. Suisse de Réassurance (1995) Les usines de traitement des déchets urbains, Zurich

51. EPA's Chemical Compatibility Chart (1980) http://www.uos.harvard.edu/ehs/environmental/EPAChemicalCompatibilityChart.pdf. Accessed 11 July 2011

52. Mallinckrodt Specialty Chemicals Co–Chemical compatibility list, 5/1989 http://www.uos.harvard.edu/ehs/environmental/MallinckrodtChemicalCompatibilityList.pdf. Accessed 29 Dec 2011

53. Cole-Palmer Instrument Company-Chemical compatibility (2011) http://www.coleparmer.com/techinfo/ChemComp.asp. Accessed 29 Dec 2011

54. University of Georgia-Chemical storage plans for laboratories (2003) http://www.esd.uga.edu/chem/chemstorage.htm, http://www.esd.uga.edu/chem/pub/lsmanual.pdf, http://www.esd.uga.edu/chem/pub/hmrelocating.pdf. Accessed 29 Dec 2011

55. The University of Vermont, http://www.uvm.edu/~esf/chemicalsafety/chemicalstorage.html. Accessed 29 Dec 2011

56. Magazine Lab Manager, Chemical storage plan fundamentals. http://www.labmanager.com/?articles.view/articleNo/1161/article/8-Chemical-Storage-Plan-Fundamentals. Accessed 29 Dec 2011

57. COMAH (Control of Major Accident Hazards), http://www.hse.gov.uk/comah/

58. Ferziger JH, Peric M (2001) Computational methods for fluid dynamics, 2nd edn. Springer, Berlin, http://elib.tu-darmstadt.de/tocs/100561322.pdf

59. Reményi K (1987) Industrial firing. Akadémiai Kiado, Budapest, 496 p

60. Ferziger JH, Peric M (2001) Computational methods for fluid dynamics, 2nd edn. Springer, New York, http://elib.tu-darmstadt.de/tocs/100561322.pdf

61. Yang YB, Nasserzadeh V, Swithenbank J (2002) Mathematical modelling of MSW incineration in a travelling bed. J Waste Manag 22(4):369–380

62. Yang YB, Goodfellow J, Nasserzadeh V, Swithenbank J (2002) Parameter study on the incineration of MSW in packed beds. J Inst Energy 75(504):66–80
63. Lim CN, Nasserzadeh V, Swithenbank J (2001) The modelling of solid mixing in waste incinerator plants. J Powder Technol 114(1):89–95
64. SUWIC papers (2011) http://www.suwic.group.shef.ac.uk/Journal%20Papers.html. Accessed 29 Dec 2011
65. Buekens A, Mertens J, Schoeters J, Steen P (1979) Experimental techniques and mathematical models in the study of waste pyrolysis and gasification. Conserv Recycl 3(1):1–23
66. Moilanen A (2006) Thermogravimetric characterisations of biomass and waste for gasification processes, VTT Publications 607. 103 pp. + app. 97 pp. Espoo, Finland
67. Nasserzadeh V, Swithenbank J, Lawrence D, Garrod N, Jones B (1995) Measuring gas-residence times in large municipal incinerators, by means of a pseudo-random binary signal tracer technique. J Inst Energy 68(476):106–120
68. Gorman P, Bergman F, Oberacker D (1984) Field experience in sampling hazardous waste incinerators. US Environmental Protection Agency, Washington, DC, EPA/600/D-84/134 (NTIS PB84201573)
69. Carroll GJ (1994) Pilot scale research on the fate of trace metals in incineration. In: Hester RE (ed) Waste incineration and the environment. Royal Society of Chemistry (Great Britain), Cambridge, pp 95–121
70. http://cfr.vlex.com/vid/270-62-hazardous-waste-incinerator-permits-19820277, (2010). Accessed 29 Dec 2011
71. Dellinger B, Torres JL, Rubey WA, Hall DL, Graham JL (1984) Determination of the thermal decomposition properties of 20 selected hazardous organic compounds. Prepared for the U.S. EPA Industrial Environmental Research Laboratory. Prepared by the University of Dayton Research Institute. EPA-600/2-84-138. NTIS PB-84-232487
72. von Paczkowski K (1979) Der Kessel als Bestandteil einer Müllverbrennungsanlage. Seine Entwicklung, sein Entwurf, WÄRME 85:121–125
73. von Paczkowski K (1984) Tendenzen bei Kesseln in Müllverbrennungsanlagen. In: Thome-Kozmiensky KI (ed) Recycling international. EF-Verlag, Berlin
74. Jachimowski A (1978) Kessel für Abfallverbrennungsanlagen. Chemie-Technik 7:403–5
75. Rasch R (1976) Korrosionsvorgänge im Feuerraum. In Kumpf, Maas, Straub, Müll und Abfallbeseitigung, E. Schmidt Verlag, 39 Lfg/III, 7300
76. Vaughan DA, Krause HH, Boyd WK (1974) Study of corrosion in municipal incinerators versus refuse composition. EPA-R-800055
77. Schroer C, Konys J (2002) Rauchgasseitige hochtemperatur-korrosion in müllverbrennungsanlagen – ergebnisse und bewertung einer literaturrecherche. Forschungszentrum Karlsruhe (FZKA), Karlsruhe, 6695
78. Brossard JM, Lebel F, Rapin C, Mareche JF, Chaucherie X, Nicol F, Vilasi M (2009) Lab-scale study on fireside superheaters corrosion in MSWI Plants. In: Proceedings of the 17th annual north american waste-to-energy conference, NAWTEC17, 18–20 May 2009, Chantilly
79. Deuerling C, Maguhn J, Nordsieck H, Benker B, Zimmermann R, Warnecke R (2009) Investigation of the mechanisms of heat exchanger corrosion in a municipal waste incineration plant by analysis of the raw gas and variation of operating parameters. Heat Trans Engin 30(10–11):822–831
80. Olie K, Vermeulen PL, Hutzinger O (1977) Chlorodibenzop-dioxins and chlorodibenzofurans are trace components of fly ash of some municipal incinerators in the Netherlands. Chemosphere 6:455–459
81. Rappe C, Andersson R, Bergqvist PA, Brohede C, Hansson M, Kjeller LO, Lindström G, Marklund S, Nygren M, Swanson SE, Tysklind M, Wiberg K (1987) Overview on environmental fate of chlorinated dioxins and dibenzofurans-sources, levels and isomeric pattern in various matrices. Chemosphere 16:1603

82. Rappe C, Andersson R, Bergqvist PA, Brohede C, Hansson M, Kjeller LO, Lindström G, Marklund S, Nygren M, Swanson SE, Tysklind M, Wiberg K (1987) Sources and relative importance of PCDD and PCDF emissions. Waste Manag Res 5(3):225–237

83. Huang H, Buekens A (1995) On the mechanisms of dioxin formation in combustion processes. Chemosphere 31:4099–4117

84. Weber R, Iino F, Imagawa T, Takeuchi M, Sakurai T, Sadakata M (2001) Formation of PCDF, PCDD, PCB, and PCN in de novo synthesis from PAH: mechanistic aspects and correlation to fluidized bed incinerators. Chemosphere 44:1429–38

85. Weber R, Sakurai T, Ueno S, Nishino J (2002) Correlation of PCDD/PCDF and CO values in a MSW incinerator–indication of memory effects in the high temperature/cooling section. Chemosphere 49:127–34

86. Sakai SI, Hayakawa K, Takatsuki H, Kawakami I (2001) Dioxin-like PCBs released from waste incineration and their deposition flux. Environ Sci Technol 35:3601–7

87. McKay G (2002) Dioxin characterisation, formation and minimisation during municipal solid waste (MSW) incineration: review. Chem Engin J 86:343–368

88. Everaert K, Baeyens J (2002) The formation and emission of dioxins in large scale thermal processes. Chemosphere 46:439–448

89. Stanmore BR (2004) The formation of dioxins in combustion systems. Combust Flame 136:398–427

90. Bumb RR, Crummett WB, Cutie SS, Gledhill JR, Hummel RH, Kagel RO, Lamparski LL, Luoma EV, Miller DL, Nestrick TJ, Shadoff LA, Stehl RH, Woods JS (1980) Trace chemistries of fire: a source of chlorinated dioxins. Science 210(4468):385–90

91. Karasek FW, Dickson LC (1987) Model studies of polychlorinated dibenzo-p-dioxin formation during municipal refuse incineration. Science 237(4816):754–756

92. Gullett BK, Bruce KR, Beach LO (1990) Formation of chlorinated organics during solid waste combustion. Waste Manag Res 8:203

93. Sidhu S, Edwards P (2002) Role of phenoxy radicals in PCDD/F formation. Int J Chem Kinet 34:531

94. Vogg H, Metzger M, Stieglitz L (1987) Recent findings on the formation and decomposition of PCDD/PCDF in municipal solid waste incineration. Waste Manag Res 5(3):285–294

95. Hagenmaier H, Kraft M, Brunner H, Haag R (1987) Catalytic effects of fly ash from waste incineration facilities on the formation and decomposition of polychlorinated dibenzo p-dioxins and polychlorinated dibenzofurans. Environ Sci Technol 21(11):1080–1084

96. Stieglitz L, Zwick G, Beck J, Roth W, Vogg H (1989) On the de-novo synthesis of PCDD/PCDF on fly ash of municipal waste incinerators. Chemosphere 18:1219–1226

97. Schwarz G, Stieglitz L (1992) Formation of organohalogen compounds in fly ash by metal-catalyzed oxidation of residual carbon. Chemosphere 25(3):277–282

98. Stieglitz L, Jay K, Hell K, Wilhelm J, Polzer J, Buekens A (2003) Investigation of the formation of polychlorodibenzodioxins /- Furans and of other organochlorine compounds in thermal industrial processes, Forschungszentrum Karlsruhe, Wissenschaftliche Berichte – FZKA 6867

99. Gullett B, Bruce K, Beach L (1990) The effect of metal catalysts on the formation of polychlorinated diobenzo-p-dioxin and polychlorinated diobenzofuran precursors. Chemosphere 20:1945–1952

100. Olie K, Addink R, Schoonenboom M (1998) Metals as catalysts during the formation and decomposition of chlorinated dioxins and furans in incineration processes. J Air Waste Manag Assoc 48:101–105

101. Kuzuhara S, Sato H, Kasai E, Nakamura T (2003) Influence of metallic chlorides on the formation of PCDD/Fs during low-temperature oxidation of carbon. Environ Sci Technol 37(11):2431–5

102. Hinton WS, Lane AM (1991) Characteristics of municipal solid waste incinerator fly ash promoting the formation of polychlorinated dioxins. Chemosphere 22:473–483

103. Tuppurainen K, Halonen I, Ruokojärvi P, Tarhanen J, Ruuskanen J (1998) Formation of PCDDs and PCDFs in municipal waste incineration and its inhibition mechanisms: a review. Chemosphere 36(7):1493–1511

104. Addink R, Paulus RHWL, Olie K (1996) Prevention of polychlorinated dibenzo-p-dioxins/dibenzofurans formation on municipal waste incinerator fly ash. Environ Sci Technol 30(7):2350–2354

105. Pandelova M, Lenoir D, Schramm K-W (2007) Inhibition of PCDD/F and PCB formation in co-combustion. J Hazard Mater 149(3):615–8

106. Vehlow J, Braun H, Horch K, Merz A, Schneider J, Stieglitz L, Vogg H (1990) Semi-technical demonstration of the 3R process. Waste Manag Res 8(6):461–472

107. Weber R, Nagai K, Nishino J, Shiraishi H, Ishida M, Takasuga T, Kondo K, Hiraoka M (2002) Effects of selected metal oxides on the dechlorination and destruction of PCDD and PCDF. Chemosphere 46:1247–1253

108. Stach J, Pekarek V, Grabic R, Lojkasek M, Pacakova V (2000) Dechlorination of polychlorinated biphenyls, dibenzo-p-dioxins and dibenzofurans on fly ash. Chemosphere 41:1881–1887

109. Alderman SL (2005) Infrared and X-ray spectroscopic studies of the copper (II) oxide mediated reactions of chlorinated aromatic precursors to PCDD/F, Ph.D. Dissertation Louisiana State University, Chapter 1. http://etd.lsu.edu/docs/available/etd-01112005-150557/unrestricted/Alderman_dis.pdf. Accessed 11 July 2011

110. Buekens A, Huang H (1998) Comparative evaluation of techniques for controlling the formation and emission of chlorinated dioxins/furans in municipal waste incineration. J Hazard Mater 62:1–33

111. Wielgosiński G (2010) The possibilities of reduction of polychlorinated dibenzo-p-dioxins and polychlorinated dibenzofurans emission. Int J Chem Eng. Review article 392175:11

112. Düwel U, Nottrodt A, Ballschmiter K (1990) Simultaneous sampling of PCDD/PCDF inside the combustion chamber and on four boiler levels of a waste incineration plant. Chemosphere 20(1):839–846, More papers are to be found at: http://www.nottrodt-ing.de/de/publi.htm

113. Wikström E, Ryan S, Touati A, Tabor D, Gullett BK (2004) Origin of carbon in polychlorinated dioxins and furans formed during sooting combustion. Environ Sci Technol 38(13):3778–84

114. Wikström E, Ryan S, Touati A, Gullett BK (2004) In situ formed soot deposit as a carbon source for polychlorinated dibenzo-p-dioxins and dibenzofurans. Environ Sci Technol 38(7):2097–101

115. Wikström E, Ryan S, Touati A, Tabor D, Gullett BK (2003) Key parameters for de novo formation of polychlorinated dibenzo-p-dioxins and dibenzofurans. Environ Sci Technol 37(9):1962–70

116. Addink R, Olie K (1995) Mechanisms of formation and destruction of polychlorinated dibenzo-p-dioxins and dibenzofurans in heterogeneous systems. Environ Sci Technol 29:1425–1435

117. Konduri R, Altwicker ER (1994) Analysis of time scales pertinent to dioxin/furan formation on fly ash surfaces in municipal solid waste incinerators. Chemosphere 28(1):23–45

118. Zimmermann R, Blumenstock M, Heger HJ, Schramm K-W, Kettrup A (2001) Emission of nonchlorinated and chlorinated aromatics in the flue gas of incineration plants during and after transient disturbances of combustion conditions: delayed emission effects. Environ Sci Technol 35:1019–1030

119. Kreisz S, Hunsinger H, Vogg H (1997) Technical plastics as PCDD/F absorbers. Chemosphere 34(5–7):1045–1052

120. Pekarek V, Weber R, Grabic R, Solcova O, Fiserova E, Syc M, Karban J (2007) Matrix effect on the de novo synthesis of polychlorinated dibenzo-p-dioxins, dibenzofurans, biphenyls and benzenes. Chemosphere (Eng) 68(1):51–61

121. Altwicker ER (1994) Formation of PCDD/F in municipal solid waste incinerators: laboratory and modeling studies. J Hazard Mater 47(1–3):137–161
122. Buekens A, Tsytsik P, Carleer R (2007) Methods for studying the de novo formation of dioxins at a laboratory scale. In: International conference on power engineering-2007, Hangzhou, 23–27 Oct 2007
123. Buekens A, Swithenbank J (2007) CFD modelling of industrial plant from a viewpoint of dioxins formation. In: International conference on power engineering (ICOPE-2007), Hangzhou
124. Verhulst V, Buekens AG, Spencer P, Eriksson G (1996) The thermodynamic behaviour of metal chlorides and sulfates under the conditions of incineration furnaces. Environ Sci Technol 30:50–56
125. http://www.termwiki.com/EN:chute-fed_incinerator_(Class_IIA). Accessed 29 Dec 2011
126. http://www.seas.columbia.edu/earth/wtert/sofos/nawtec/1964-National-Incinerator-Conference/1964-National-Incinerator-Conference-25.pdf. Accessed 29 Dec 2011
127. http://www.dioxinfacts.org/sources_trends/trash_burning.html. Accessed 29 Dec 2011
128. http://www.epa.gov/oaqps001/community/details/barrelburn.html. Accessed 29 Dec 2011
129. Gullett BK, Lemieux PM, Lutes CC, Winterrowd CK, Winters DL (1999) PCDD/F emissions from uncontrolled, domestic waste burning. Presented at Dioxin '99, the 19th international symposium on halogenated environmental organic pollutants and POPs, Organohalogen compounds, vol 41, Venice, 12–17 Sept 1999, pp 27–30
130. Lemieux PM, Lutes CC, Abbott JA, Aldous KM (2000) Emissions of polychlorinated dibenzo-p-dioxins and polychlorinated dibenzofurans from the open burning of household waste in Barrels. Environ Sci Technol 34:377–884
131. iran
132. http://www.sittommi.fr/fonctionnement-usine-incineration-ordures-menageres-pontivy.html. Accessed 29 Dec 2011
133. Buekens A, Yan M, Jiang XG, Li XD, Lu SY, Chi Y, Yan JH, Cen K (2010) Operation of a municipal solid waste incinerator – Pontivy. i-CIPEC
134. Winnacker
135. Saxena SC, Jotshi CK (1994) Fluidized-bed incineration of waste materials. Prog Energy Combust Sci 20(4):281–324
136. Integrated pollution prevention and control reference document on best available techniques for the waste treatments industries, August 2006
137. Santoleri JJ (1972) Chlorinated hydrocarbon waste recovery and pollution Abatement. In: Proceedings of the 1972-National-incinerator-conference, New York
138. Mizuno K (2002) Destruction Technologies for ozone depleting substances in Japan. National Institute for Resources and Environment, in UNEP: http://www.unep.fr/ozonaction/information/mmcfiles/3521-e-file2.pdf. UNON Nairobi
139. http://www.uneptie.org/ozonaction/information/mmcfiles/3521-e-file2.pdf. Accessed 29 Dec 2011
140. http://submergedcombustion.org.uk/Default.aspx. Accessed 29 Dec 2011
141. Tsukishima Kankyo Engineering (2010) http://www.tske.co.jp/english/index.html. Accessed Dec 2011
142. Buekens AG, Schoeters JG, Jackson DV, Whalley LW (1986) Status of RDF-production and utilization in Europe. Conserv Recycl 9:309–309
143. Friends of the Earth (2008) Briefing – mechanical and biological treatment (MBT)
144. Wikipedia, Mechanical and biological treatment (MBT)
145. IPPC (1999) Integrated pollution prevention and control (IPPC): reference document on best available techniques in the cement and lime manufacturing industrie. Formation and release of POPs in the cement industry, 2nd edn. European Commission, Directorate General JRC, Institute for Prospective Technological Studies, Seville
146. Ökopol (1999) Economic evaluation of dust abatement techniques in the European cement industry, Report for EC DG11, contract B4-3040/98/000725/MAR/E1; and "Economic

evaluation of NOx abatement techniques in the European cement industry", Report for EC DG11, contract B4-3040/98/000232/MAR/E1. Ökopol GmbH, Hamburg

147. Rabl A (2000) Criteria for limits on the emission of dust from cement kilns that burn waste as fuel. ARMINES/Ecole des Mines de Paris, Paris

148. SINTEF (2006) Formation and release of POPs in the cement industry, second edition. Report of the World Business Council for Sustainable Industry, Cement sustainability initiative, Geneva

149. Greenpeace International (1991) http://archive.greenpeace.org/toxics/reports/gopher-reports/inciner.txt. Amsterdam. Accessed 29 Dec 2011

150. Greenpeace International (1994) http://archive.greenpeace.org/toxics/reports/azd/azd.html. Greenpeace Communications, London

151. Costner P (2001) Chlorine, Combustion and Dioxins: Does Reducing Chlorine in Wastes Decrease Dioxin Formation in Waste Incinerators? http://archive.greenpeace.org/toxics/reports/chlorineindioxinout.pdf

152. PVC WASTE AND RECYCLING. Solving a Problem or Selling a Poison? (1999) http://archive.greenpeace.org/toxics/html/content/pvc3.html#top. Accessed 29 Dec 2011

153. Buekens A, Cen KF (2011) Waste incineration, PVC, and dioxins. J Mater Cycles Waste Manag 13:190–197

154. Xu MX

155. Travis CC (1991) Municipal waste incineration risk assessment: deposition, food chain impacts, uncertainty, and research needs. Plenum Press, New York

156. Hattemer-Frey HA, Travis CC (1991) Health effects of municipal waste incineration. CRC Press, Baco Raton

157. Roberts SM, Teaf CM, Bean JA (1999) Hazardous waste incineration: evaluating the human health and environmental risks. Lewis, Boca Raton

Books and Reviews

Air Pollution Control Association, American Society of Mechanical Engineers. Research Committee on Industrial and Municipal Wastes (1988) Hazardous waste incineration: a re-source document sponsored by the ASME Research Committee on Industrial and Municipal Wastes; co-sponsored by the Air Pollution Control Association, the American Institute of Chemical Engineers, the United States Environmental Protection Agency

Alter H, Horowitz E (1975) STP 592, Resource recovery and utilization. In: Proceedings of the national materials conservation symposium. http://www.astm.org/BOOKSTORE/PUBS/STP592.htm. Accessed July 2011

Bilitewski B, Härdtle G, Marek K (2000) Abfallwirtschaft. Handbuch für Praxis und Lehre. Springer, Berlin

Bonner T, Dillon AP (1981) Hazardous waste incineration engineering, pollution technolo-gy review 88. Noyes Data Corporation, Park Ridge

Gershman, Brickner & Bratton, Inc. (1986) Small-scale municipal solid waste energy recovery systems. Van Nostrand Reinhold, New York

de Souza-Santos ML (2004) Solid Fuels combustion and gasification: modeling, simulation, and equipment operations. Marcel Dekker, New York

Freeman HM (1988) Incinerating hazardous wastes. Technomic, Lancaster

Görner K (1991) Technische Verbrennungssysteme. Springer, Berlin\Heidelberg\New York

Grover VI (2002) Recovering energy from waste: various aspects. Science, Enfield

Günther R (1974) Verbrennung und Feuerungen. Springer, Berlin\Heidelberg\New York

Hester RE (1994) Waste incineration and the environment. Royal Society of Chemistry (Great Britain), Cambridge

Institute of Electrical and Electronics Engineers (1975) Incineration and treatment of hazardous waste. In: Proceedings of the eighth annual research symposium CRE: conversion of refuse to energy, vol 1. World Environment and Resources Council, Institute of Electrical and Electronics Engineers

International conference on combustion, incineration/pyrolysis (i-CIPEC). In: Proceedings of the 1st (Seoul, Korea in 2000), 2nd (Jeju, Korea in 2002), 3rd (Hangzhou, China in 2004), 4th (Kyoto, Japan in 2006), 5th (Chiangmai, Thailand in 2008), and 6th International Conference on Combustion, Incineration/Pyrolysis (Kuala Lumpur, Malaysia, 2010)

International conference on thermal treatment technologies

National Research Council (US). Committee on Health Effects of Waste Incineration (2000) Waste incineration and public health. National Academies, Washington

National-Incinerator-Conference 1964, 1966, 1968, 1970, 1972, 1974 (visit the proceedings at the WTERT-site of Columbia University, e.g. at http://www.seas.columbia.edu/earth/wtert/sofos/nawtec/1966-National-Incinerator-Conference/). Accessed July 2011

National-Waste-Processing-Conference 1976, 1978, 1980, 1982, 1984 1986, 1988, 1990, 1992, 1994 (visit the proceedings at the WTERT-site of Columbia University, e.g., http://www.seas.columbia.edu/earth/wtert/sofos/nawtec/1980-National-Waste-Processing-Conference/). Accessed July 2011

North American Waste to Energy Conferences (NAWTEC) http://nawtec.swana.org/. Accessed July 2011

EPA (1989) Environment Canada. Proceedings of the international conference on municipal waste combustion. Hollywood, Florida

Robinson WD (1986) The solid waste handbook: a practical guide. Wiley, Chichester

Rogoff MJ, Screve F (2011) Waste-to-energy: technologies and project implementation. Elsevier Science, Amsterdam

Santoleri JJ, Theodore L, Reynolds J (2000) Introduction to hazardous waste incineration. Wiley-IEEE, New York

Solid Waste Association of North America (1998) Asian-North American solid waste management conference. Paper presented at the 17th biennial waste processing conference, Atlantic City (Proceedings available at the WTERT-site of Columbia University)

Theodore L, Reynolds J (1987) Introduction to hazardous waste incineration. Wiley, New York

Warnatz J, Maas U, Dibble RW (2001) Combustion – physical and chemical fundamentals, modeling and simulation, experiments, pollutant formation, 3rd edn. Springer, Berlin\Heidelberg\New York

World Health Organization. Regional Office for Europe (1985) Solid waste management: selected topics. World Health Organization, Copenhagen

Young GC (2010) Municipal solid waste to energy conversion processes: economic, technical, and renewable comparisons. Wiley, Hoboken